The Man Who Was in Two Places at Once

A Modern Tale of Super-powers

R.A. Varghese

Testament Book House
PMB 19611
228 Park Ave S
New York, NY 10003

ISBN 978-1-0880-4947-1

Published September 2022

A gift to our spiritual father from the spiritual children of Padre Pio

"I have made a pact with the Lord: I will take my place at the gate to paradise, but I shall not enter until I have seen the last of my spiritual children enter."
— Padre Pio of Pietrelcina

Table of Contents

Perfect Storm

Moderners pride themselves on being too sophisticated to fall for religion, the superstition embraced by the simpletons of earlier times. After all, the "assured results" of science had shown us that religious beliefs are fanciful fairy tales. And the first astronaut, Yuri Gagarin, had dutifully reported in 1961 that he did not find any deity in outer space. No rational person could thereafter respectably believe in anything outside the natural. And miracles were obviously impossible since they would violate the immutable laws of nature.

As if to confound these pompous pretensions, Heaven played a joke on the self-proclaimed rationalists who propounded them. The heavenly humor came in the person of a simple, earthy monk who was dragged, virtually kicking and screaming, into the global limelight.

Like a meteor from the heavens, this monk from nowhere tore through the veil of skepticism and scientism erected by his contemporaries. In his very physical being and his everyday activity, he was a stupefying witness to a world beyond the here-and-now. In the most natural way possible, he was an astounding testimony to the supernatural.

Yet, paradoxically, at another level he personally manifested some of the profoundly puzzling ideas being unveiled by science itself. For the science of Relativity and Quantum Physics and parallel universes and the like was distant from the dogmas of the drumbeaters of scientism but not from the life and times of the monk from Pietrelcina.

Padre Pio, as the monk was affectionately known, ended up being recognized as one of the most remarkable saints in history. Saints are canonized on the basis of demonstrable sanctity while on earth and at least two verifiable miracles attributable to their posthumous intercession. Depending on how you look at it, the Padre was either a unique kind of saint or every kind of saint. Each of the multitudes of canonized saints usually had one well-known charism or gift – a "super-power" of sorts granted by God. Padre Pio, St. Pio of Pietrelcina, is reputed to have had the charisms of *all* the saints.

Consider this sampling taken from among thousands of reports:

- He could bilocate, i.e., be in two places at the same time.
- He also seemed to travel to the future and back.
- He suffered the stigmata, the wounds of Christ on his body. The wounds would start bleeding at the moment of the consecration event at Mass.
- He was known for the spiritual and physical healings attributed to his intercession.
- He could read minds.
- He used to physically wrestle with the Devil.
- He "flew" in the air as witnessed by various pilots including those he saved after being ejected without parachutes.
- Through his intercession, many souls were freed from the state of cleansing that comes before Heaven and came and conversed with him in gratitude.
- He had the gift of fragrance – you could know that he was near you from a scent of roses or something else associated with him.
- He took on people as his spiritual children – a ministry unique to him – and interceded in their lives when they asked.

Padre Pio was truly a fitting saint for the shroud of skepticism known as the Modern Age.

Faced with the phenomenon that was Padre Pio, the prophets of scientism (of the Victorian variety) predictably scurried around for ways in which they could discredit him. His stigmata were generated by carbolic acid. His miracles were simply legends concocted by his illiterate followers. The accounts of his bilocation were unverifiable hearsay.

But the Padre's antagonists were not simply non-believers. Right from the start, clerics for whom the Church was a power play were not amused by his call to radical conversion. Ecclesial authorities did not welcome what they saw as competition given his popularity with the masses. He was rocking the boat – their boat! Like the "rationalists", the ecclesiasts denounced the Padre as a fraud and diligently sought to expose him.

If the skeptics were Sadducees who trafficked in ridicule, the ecclesiastics were Pharisees who specialized in entrapment. But neither succeeded for long. The Padre had opened a window into a different order of reality. And too many people had peered through this window for armchair authorities to airily dismiss its existence.

Like Chico Marx, the foes were reducing to saying, "Who are you going to believe? Me or your lying eyes?"

In retrospect, the positively hostile attitude of highly placed clerical officials turned out to be a good thing. There could be no plausible argument that the Padre's miracles were "creations" of the Church when the very same Church regularly silenced him and forbade him from all contact with the public.

The Devil's Advocate, usually deployed against candidates for canonization long after their deaths, was unleashed on him from the start of his public life.

So was the Devil himself! Padre Pio had both feet in the natural and the supernatural realms at the same time. He had to deal with the denizens of both realms – good and bad – 24/7.

The inevitable question at this point is: how do we know that the Padre actually did do all that was claimed about him? This leads us to a more fundamental question: How do we know ANYTHING?

Some things we know from the evidence of our senses. For instance, if we step outside on a summer's day and say that the sun is shining we are speaking about something we see. There are other things that cannot be proved via the senses or logic but that we KNOW to be true since they are self-evident. For instance, we know that we are conscious, i.e., we are aware of our environment, and our ability to be thus aware distinguishes us from a rock. We cannot prove that we are conscious because every proof would have to assume our being conscious but we cannot deny it either – unless we were simply insane. This is why we call it a self-evident truth.

For the most part, Padre Pio's miracles involved the senses. Normally, we organize what we know through our senses under the rubric of science. We construct theories and verify them by finding observable evidence in their favor. A claim is considered scientific if it makes predictions that can be verified by repeatable experiments. A drug is considered effective if it can be shown to have a beneficial effect on large numbers of people suffering from the same ailment. Universality, repeatability and verifiability are essential hallmarks of valid scientific claims.

So how do these criteria apply with respect to the Padre Pio phenomenon? They don't. This is because we are not talking about scientific theories or laws here but about encounters, experiences, observations and testimonies of witnesses. We are talking about history not physics, a court case and not a measurement of quantities. The evidence assessment has to be done in the manner of law courts and not laboratories.

Astronomical observations, clinical trials and lab tests are not the only ways of evaluating evidence. In a court of law, one or more eye-witnesses are sufficient to provide evidence that counts as "direct" and not "circumstantial". Cases can be built using fingerprints, DNA, motive, evidence that is direct or circumstantial, witnesses. No one criticizes this procedure as being "anecdotal" or "unscientific" because they recognize it as the only applicable approach to the evidence involved. As the court system puts it:

> Evidence may be direct or circumstantial. Direct evidence is direct proof of a fact, *such as testimony by a witness about what that witness personally saw or heard or did.* Circumstantial evidence is indirect evidence, that is, it is proof of one or more facts from which one can find another fact.
>
> You are to consider both direct and circumstantial evidence. Either can be used to prove any fact. The law makes no distinction between the weight to be given to either direct or circumstantial

evidence. It is for you to decide how much weight to give to any evidence.[1]

If thousands of people in recent times claim to have witnessed Padre Pio's stigmata, complemented by film footage and other documentary evidence, then this claim has to be assessed like any other claim made in a court of law. For instance, we were informed by both eye-witness and documentary testimonies that thousands of people perished in the Nazi gas chambers. We hold this claim to be true based on these testimonies as well as various other kinds of evidence (the discovery of the remains of the victims, the confessions of the perpetrators, film footage of victims and survivors).

In like manner, the claims about the prodigies of Padre Pio have to be assessed by assembling the various kinds of evidence with which we are presented. These are primarily the testimonies of those who have witnessed or experienced what is claimed. One person may have witnessed the stigmata while another might have experienced an inexplicable healing. And these experiences continue to the present day. We also have film footage, the testimony of medical specialists, and other kinds of evidence. The variety and volume of the data thus assembled make up a cumulative case for the reality of Padre Pio's parallel universe.

Of course, the Padre Pio prodigies are not repeatable and verifiable by us in the manner of standard scientific claims or laws. But then again science has moved on since the Victorian Age although many of its drumbeaters still have Victorian mindsets. Not only have we witnessed the revolutions of relativity and quantum theory but scientists today matter-of-factly make claims that they themselves say can never be verified.

Take the currently popular claim of the multiverse. This is the claim that there is an infinite number of parallel universes generated from quantum energy fields, black holes, et al. There are no immutable laws of physics that apply in every universe. Instead, there are specific laws for each. And each of these universes are "shut off" from all the others so there is no way to verify their existence.

The multiverse is just one of several such claims in contemporary science. In his analysis of current theoretical physics, *Farewell to Reality*, Jim Baggott writes, "There is as yet no observational or experimental evidence for many of the concepts of contemporary theoretical physics, such as super-symmetric particles, superstrings, the multiverse, the universe as information, the holographic principle or the anthropic cosmological principle. For some of the wilder speculations of the theorists there can by definition never be any such evidence."[2]

The key takeaway here is this: today's scientists say that the laws of nature are not immutable. Physical laws that apply in our universe do not apply in an adjoining parallel universe. And this claim cannot be disproven – or proven – using current models.

In this context, the idea that Padre Pio was stepping in and out of a parallel universe of a different kind does not sound "unscientific" at least in terms of present-day science.

Yes, this parallel universe, as understood by the Padre's afficionados, followed laws that went beyond the physical order. Such an affirmation can hardly be considered dubious in itself from the standpoint of contemporary culture. The existence of the non-physical is taken for granted by even the non-religious population. The perennial popularity of TV shows like *Paranormal, Stranger Things* and *Supernatural* and of the horror and exorcist movie genres testifies to the public's instinctive belief in the non-physical. In fact, according to a 2018 Chapman University survey, "More than three quarters of Americans believe in at least one paranormal phenomenon."[3]

What is striking about the Padre's "other" universe is that we are not dealing with something esoteric, obscure and fragmentary. Padre Pio (to his chagrin) was operating under the public eye. The remarkable phenomena associated with him were as frequent as they were diverse.

He was a gift that kept on giving. And, in fact, continues to keep on giving. For, as the Padre's spiritual children will testify, the Padre Pio phenomenon continues to manifest itself to this very day.

In this work, we will consider the prodigies attributed to Padre Pio in the context of analogous ideas and claims from modern science. We talk of an electron being in two places at once. This being the case, we can likewise explore the idea of a man, specifically Padre Pio, being in two places at once.

Notes

[1]https://www.ce9.uscourts.gov/jury-instructions/node/304
[2]Jim Baggot, *Farewell to Reality* (New York: Pegasus, 2013), ix, x, 292.
[3]https://blogs.chapman.edu/wilkinson/2018/10/16/paranormal-america-2018/

The Magical World
of Modern Science

Before we consider the life and times of Padre Pio, let us enter into the magical world of modern science. We say magical because the world unveiled by the science of our day is more mysterious than anything our most primitive ancestors could have conjured up.

Consider these headlines from major scientific publications:

- "2,000 Atoms Exist in Two Places at Once in Unprecedented Quantum Experiment"[1]
- "Einstein's "Time Dilation" Prediction Verified – a person travelling in a high-speed rocket would age more slowly than people back on Earth"[2]
- "The Case for Parallel Universes – Why the multiverse, crazy as it sounds, is a solid scientific idea"[3]

These articles were published in the world's leading scientific journals: the first headline concerns a study published in *Nature* and the second and the third come from *Scientific American.*

From quantum physics, we learn of the strange phenomenon of "quantum super-positioning" where a quantum system exists in various quantum states at the same time. From relativity we learn that space-time is curved by large bodies and gravity arises from the shape taken by space-time.

As for the origin of the universe, we learn that

> Around 13.8 billion years ago, all the matter in the Universe emerged from a single, minute point, or singularity, in a violent burst. This expanded at an astonishingly high rate and temperature, doubling in size every 10-34 seconds, creating space as it rapidly inflated. Within a tiny fraction of a second gravity and all the other forces were formed.[4]

Amazing as all this might be, the scientific enterprise has even greater surprises in store.

We learn that everything that exists – atoms and cells, energy fields and ecosystems – is constantly *processing information*. Everything – from photons and planets to microbes and mammals – is constantly *moving to a goal*, advancing from one state to another (Photons, the particles that make up light, travel over five trillion miles a year and have been doing this for billions of years). Everything – from particles to butterflies – is constantly *causing effects* in its environment. Everything is a manifestation of *mind*. This is the message of modern science.

The new story of science also tells us that everything around us belongs to a social network. These social networks are created by mathematically structured laws of nature and the actions of various kinds of agents.

- Electrons and quarks belong to the social network of atoms (a grain of sand has over 50 quintillion atoms)
- molecules are networks of atoms
- stars belong to galactic social networks (there are at least 200 billion trillion stars in the universe)
- microbes exist in dynamic, complex communities called microbiomes (some 100 trillion microbes live either within or on the outside of each of us)
- fungi create the wood-wide web that enables the global cycling of nutrients
- plants belong to a social network of sun, soil and bees
- ants create anthills, birds fly in flocks

- animals operate in herds
- each human being is made up of 38 trillion constantly interacting cells

All these social networks have their own kinds of messaging platforms. An electron's tweet is a photon released when it changes orbits within an atom. Bacteria put up posts on their social networks by releasing specific chemical molecules. The posts enable them to coordinate attacks on targets.

Plants are constantly podcasting, livestreaming and tweeting using chemical signals via air or soil. They warn each other of impending attacks, say, from predatory pests. Some use forms of mimicry, others manipulate insects with neuroactive substances. Insects communicate through emissions of sound and light and chemical odor. Their direct messaging serves many purposes: reproduction, alerts, defense and more.

Birds use acoustic communication for courting or warning or even mobilizing their fellows for an attack. Bird e-commerce takes the form of migration. Many species of birds migrate to places thousands of miles away during certain periods of the year to secure food or nesting locations. Fish communicate through sound and color and some use electrical signaling. Fish social networks help individual fish locate food, evade prey or mate.

Non-human mammals communicate through visual, auditory and chemical messaging platforms. They use these to find food and mates and to fight or flee from predators. Herds have their own versions of relationship and group messaging networks through which they cooperate and coordinate.

The cosmos itself is a network of networks. It is a metaverse of mind.

But how did this metaverse of mind come to be?

We can find a clue to its origin by studying the part of the metaverse for which we have a user login – ourselves.

Our brains are made up of some 80 billion neurons. Each neuron is made up of an average of some 100,000 molecules. These molecules change thousands of times over the course of our lives. And yet we remain the same at two months old, two years old, 20 years old and all the way to now.

Of course, we grow and learn and change but it is the same "I" and "you" that undergoes these changes.

So what is it that stays the same? It is clearly not something physical because the physical – take, for instance, our neurons – is in a constant state of flux. We could not even complete a sentence if we were entirely physical. *Clearly there is something about us that is non-physical.*

Where did this come from? Not from the physical for two reasons: the physical is constantly changing, not just the neurons in our brains but all our other cells, and, secondly, the laws of physics have no room for the creation of the non-physical from the physical.

So the non-physical "I" and "you" must come from something that is itself non-physical. And where did this non-physical "something" come from? It must always have existed without limitation. *It is Mind infinite and eternal.*

As Georg Cantor, the father of the modern theory of mathematical infinity, put it:

> "What surpasses all that is finite and transfinite is no 'Genus'; it is the single, completely individual unity in which everything is included, which includes the Absolute, incomprehensible to the human understanding. This is the Actus Purissimus, which by many is called God."[5]

This is why Max Planck, the father of quantum physics, said:

> "All matter originates and exists only by virtue of a force which brings the particle of an atom to vibration and holds this most minute solar system of the atom together. We must assume behind this force the existence of a conscious and intelligent mind. This mind is the matrix of all matter."[6]

Likewise, Albert Einstein, the greatest scientist of all time, said:

> "A conviction … of the rationality or intelligibility of the world lies behind all scientific work of a higher order … a belief bound up with deep feeling, in a superior mind that reveals itself in the world of experience."[7]

Charles Darwin, the father of modern evolutionary theory, observed "I feel compelled to look to a First Cause having an intelligent mind in some degree analogous to that of man."[8]

Charles Scott Sherrington, the creator of modern neurophysiology, rightly said, "As followers of natural science we know nothing of any relation between thoughts and the brain, except as a gross correlation in time and space. (…) Biology cannot go far in its subject without being met by mind."[9]

The Nobel Prize winning biologist George Wald added, "Mind, rather than emerging as a late outgrowth in the evolution of life, has existed always as the matrix, the source and condition of physical reality (…) It is mind that has composed a physical universe that breeds life."[10]

It is clear then that the world revealed by modern science is mysterious beyond anything conceived by Enlightenment "rationalists" like David Hume and today's neo-Victorian atheists. Given this new scientific appreciation for the underlying metaverse of mind, it is time to take a fresh look at a human phenomenon that was is mysterious as any of the new narratives of modern science – Padre Pio of Pietrelcina.

Notes

[1] https://www.space.com/2000-atoms-in-two-places-at-once.html
[2] https://www.scientificamerican.com/article/einsteins-time-dilation-prediction-verified/
[3] https://www.scientificamerican.com/article/multiverse-the-case-for-parallel-universe/
[4] https://www.bbc.com/future/article/20140812-how-was-the-universe-created

[5]Joseph Dauben, *Georg Cantor: His Mathematics and Philosophy of the Infinite*, (Princeton: Princeton University Press, 1990): 290.

[6]Max Planck, *Das Wesen der Materie* [The Nature of Matter], speech at Florence, Italy (1944) (from Archiv zur Geschichte der Max-Planck-Gesellschaft, Abt. Va, Rep. 11 Planck, Nr. 1797).

[7]Albert Einstein, *Ideas and Opinions*, trans. Sonja Bargmann (New York: Dell Publishing Company, 1973), 255.

[8]Charles Darwin, *The Autobiography of Charles Darwin*, https://www.update.uu.se/~fbendz/library/cd_relig.htm.

[9]Charles Sherrington, *Man on his Nature* (New York: The Macmillan Company, 1941), 290-291.

[10]George Wald, 1986 address to the First World Congress for the *Synthesis of Science and Religion*.

The Man Who Loved Too Much

"For God so loved the world that he gave his only Son, so that everyone who believes in him might not perish but might have eternal life." John 3:16

"No one has greater love than this, to lay down one's life for one's friends." John 15:13

"Then Jesus said to his disciples, 'Whoever wishes to come after me must deny himself, take up his cross, and follow me.'" Matthew 16:24

"I rejoice in my sufferings for your sake, and in my flesh I am filling up what is lacking in the afflictions of Christ on behalf of his body, which is the church." Colossians 1:24

"'These signs will accompany those who believe: in my name they will drive out demons, they will speak new languages. They will pick up serpents [with their hands], and if they drink any deadly thing, it will not harm them. They will lay hands on the sick, and they will recover.'" Mark 16:17-8

Man on a Mission

Born on May 25, 1887, Francesco Forgione, later known as Padre Pio, was the son of a peasant couple from the southern Italian village of Pietrelcina in the province of Benevento. At the age of five he committed his life to God accompanied by a regimen of penance and devotion. He worked at his family farm until the age of ten. He had already decided that he was going to become a priest and at the age of 15 joined the Franciscan Capuchin Order of the Friars Minor. He was ordained a priest in 1910.

Francesco, who took the name Pio upon becoming a friar and was called Padre Pio after ordination, witnessed various supernatural manifestations from an early age. These included attacks of what he perceived to be an evil force as well as encounters with the angels, Jesus and Mary. He also fell ill frequently and was eventually transferred to the Capuchin monastery in San Giovanni Rotondo. In 1915, he felt a sharp pain in his arms and feet. These came from wounds that left no marks and were called invisible stigmata. Stigmata, as traditionally understood, refer to bodily wounds that correspond to the wounds suffered by Jesus on the cross. Padre Pio was soon to have a visible version of the stigmata.

In 1918 things moved to a new level as Bert Ghezzi narrates:

> A series of mystical events that occurred in August and September of 1918 shaped the remainder of Padre Pio's life. On August 5, an angel like the one who struck Teresa of Ávila's heart appeared to him and wounded his soul. Later he described to his spiritual director what had happened: "I was filled with extreme terror at the sight of a heavenly Being who presented himself to the eye of my intellect. In his hand he held some kind of weapon, like a long, sharp-pointed steel blade, which seemed to spew out fire. He hurled this weapon into my soul with all his might. It was only with difficulty that I did not cry out. I thought I was dying. ... This agony lasted until the morning of August 7.... Even my internal organs were torn and ruptured by that weapon. Since that day I have been mortally wounded. I always feel in the depths of my soul a wound that is always open and that causes me continual agony."
>
> Then after celebrating Mass on September 18, while sitting in the monastery choir before an ancient crucifix, Padre Pio received the stigmata. He says he fell into a deep, peaceful repose, and an angel with blood oozing from his hands, feet, and side appeared to him. "I felt like I was dying," he said, "and I would have died if the Lord had not intervened to strengthen my heart, which was ready to burst out of my chest. When the mysterious creature left, I found that my hands, feet, and side had been pierced and were bleeding. ... The wound in my heart bleeds continuously, especially from

Thursday evening until Saturday. I'm afraid that I will bleed to death if the Lord doesn't hear my groans and take these wounds from me. He can even leave the anguish and the pain, but let him take away these visible signs that are a source of embarrassment for me and an indescribable and unbearable humiliation."

For his own good reasons, the Lord said no to Padre Pio's prayer.[1]

It soon became clear to Padre Pio that his mission was to suffer a living martyrdom for the salvation and sanctification of his contemporaries and the liberation of those who were being cleansed in the antechamber of Heaven. Italian journalist Andrea Tornielli writes:

> The desire to suffer in order to ransom the souls of sinners was constant throughout his whole life. He himself, on November 29, 1910, had asked in writing to Padre Benedetto of San Marco in Lamis for special permission in this regard: "For some time past I have felt the need to offer myself to the Lord as a victim for poor sinners and for the souls in Purgatory. This desire has been growing continually in my heart, so that it has now become what I would call a strong passion. I have in fact made this offering to the Lord several times, beseeching him to pour out upon me the punishments prepared for sinners and for the souls in a state of purgation, even increasing them a hundredfold for me, as long as he converts and saves sinners and quickly admits to paradise the souls in Purgatory, but I should now like to make this offering to the Lord in obedience to you."

> Padre Benedetto understood, and gave his consent. From that time the saint of the Gargano, who had begun to experience in himself the pains of the stigmata, always renewed this total donation. "No, I want to suffer until the end of the world," he replied one day to Malvina Lureti, who had advised him to take some rest. This is also confirmed by Padre Pierino Galeoni, a priest who was a spiritual child of Padre Pio: "He revealed to me that he asked Jesus to be able to be not only a perfect victim, but also a perennial victim, that is to continue to remain a victim in his children (…).

He confirmed to me that he had obtained from the Lord the mission of being victim and father of victims until the last day."[2]

In addition to suffering the constant pain and embarrassment of the stigmata, Padre Pio carried out his mission through the confessional, daily Mass, and spiritual direction for the faithful.

> Padre Pio spent most of his day hearing confessions. From 1918 to 1923, he heard confessions fifteen to nineteen hours a day. During the 1940's and the 1950's, about eight hours a day. In 1962, 83,035 women and 19,837 men registered for confession with Padre Pio: an average of about 273 per day. In 1967, Padre Pio heard confession of about 15,000 women and 10,000 men: an average of 70 people per day.

> People would come to him from all over the world. He heard the confessions in the old church, even after the new church was built.[3]

He was a modern-era Job, a living catechism. In his mission he embodied Heaven's answer to the problem of suffering. The divine response to the pain and suffering caused by evil was the sacrificial salvific death of the Man of Sorrows on Calvary and the freely-offered participation of his followers in that once-for-all sacrifice.

We should also consider Padre Pio in terms of the bigger picture of the workings of Providence. On October 13, 1884, Pope Leo XIII was shown in a vision that the Devil would have great power for a period of 100 years. In response, the Pope wrote a prayer to St. Michael to be recited at all churches. And, in 1899, in union with all the bishops of the Church, he consecrated the world in the next century to the Holy Spirit.

Heaven answered decisively. 33 years to the day after Pope Leo's vision there was the great miracle of the sun at Fatima. Fatima, which was tied to all the momentous events of modern times, was followed by other major Marian apparitions. At a time when evil seemed more triumphant than ever, there was the Divine Mercy revelation of Jesus to Sister Faustina Kowalska. On a different plane, we had the extraordinary ministries of Mother Teresa of Calcutta and Pope John Paul II, two contemporary saints whose lives had an impact not simply on the Church but on the world. And then there was Padre Pio who became a living witness to the supernatural and who lived a stone's throw away from Mount St. Michael, one of only three places where the Archangel Michael is said to have appeared to the faithful.

The Empires Strike Back

The phenomenon of Padre Pio's stigmata was a hard fact that mesmerized his contemporaries, infuriated Church authorities and sent atheists into orbit. The race was on now to discredit him and expose the phenomenon as a fraud.

Church experts dispatched to examine him hypothesized that the stigmata had a purely natural origin. They were caused either by chemical agents like carbolic acid or generated by his unstable mind because of its obsession with the suffering of Jesus. Clerics – from the Twenties to the Sixties! – also sought to "dig up dirt" on the Padre by planting tape recorders in the confessional, assembling salacious gossip, fabricating stories and soliciting critiques from monks jealous of him.

The chemical and psychosomatic critiques of the stigmata fizzled out quite early in the game. The carbolic acid charge was made by theologians and journalists who knew nothing about chemistry. Although carbolic acid had some use as a disinfectant, it is now known to be toxic in high doses. If indeed it had been applied to create wounds (as alleged), the effect would have been fatal. The US Centers for Disease Control warns that if it "is left on the skin, it will penetrate rapidly and lead to cell death and gangrene."[4]

As for the psychosomatic origin idea, this is still the go-to hypothesis of stigmata skeptics from the 19th century to the present. To be sure, there are "hysterics" who seem to make certain kinds of bloody marks appear of their own volition; some appear to have bloody sweat under stress; others have generated them under hypnosis. But these are superficial, temporary and haphazard bloody markings on the skin of various kinds and shapes and in different locations. Those afflicted with them are often mentally unbalanced. And, of course, there are fraudsters who have inflicted wounds on themselves and claimed to be stigmatists.

But, as with apparitions and exorcisms, we should first look for all "natural" explanations in any claim of stigmata. Yes, some claims are based on fraud. Others could be wounds that appear from mental suggestion or an altered state of mind (although this route remains a hypothesis). But in certain cases, it is obvious that a psychosomatic explanation is simply implausible. Fr. Benedict Groeschel, who examined "the nature of reported supernatural phenomena and consider[ed] the different possibilities ranging from the psychosomatic to the miraculous," distinguished Padre Pio's stigmata from the markings found on others whom he had examined: "I don't think any of the several stigmatics I have examined exhibited supernatural markings; in fact, I was able to help one devout and sincere person recover from this unusual symptom. Uniformly these wounds were superficial—something like blood blisters. Padre Pio's wounds were deep, and they bled constantly but without any infection."[5]

Each claim of the supernatural should be examined on its own merits.

Authentic Marian apparitions often leave behind enduring "signs" that can be examined today as, for instance, the scientifically inexplicable Tilma of Guadalupe or the healing springs of Lourdes. The water at Lourdes has no curative or medicinal properties but is known for its healing power. Oxford University's *Journal of the History of Medicine and Allied Sciences* noted that "the Lourdes phenomenon, extraordinary in many respects, still awaits scientific explanation."[6] Yet it is little known that, as the Lourdes apparitions gained acceptance, over thirty claims of other apparitions sprang up in surrounding areas. But none of these had any supporting evidence and were later found to be fraudulent.

The point is that counterfeit currency notes do not disprove the existence of real currency. Quite the opposite!

So how do you distinguish genuine stigmata from self-deception and outright deception?

In authentic stigmatization, writes John Hardon," the wounds are localized in the very spots where Christ received the five wounds, which does not occur if the bloody sweat is produced by hysteria or hypnotism. Generally the wounds bleed afresh and the pains recur on the days or during the seasons associated with the Savior's passion, such as Fridays or feast days of Our Lord. The wounds do not become festered and the blood flowing from them is pure, whereas the slightest natural lesion in some other part of the body develops an infection. Moreover, the wounds do not yield to the usual medical treatment and may remain for as long as thirty to forty years. The wounds bleed freely and produce a veritable hemorrhage; and this takes place not only at the beginning but again and again. Also the extent of the hemorrhage is phenomenal; the stigmata lie on the surface, removed from the great blood vessels, yet the blood literally streams from them. Finally true stigmata are not found except in persons who practice the most heroic virtues and possess a special love of the Cross."[7]

Honest skeptics, while rejecting the reality of this phenomenon, admit the limits of what we know and can know about stigmata. Benjamin Medford, the deputy editor of *Skeptical Inquirer*, writes "If stigmata is real, there is no medical or scientific explanation for it. Wounds do not suddenly and spontaneously appear on people's bodies for no reason."[8] An article on the veracity of stigmata claims in *The Smithsonian* notes, "The jury remains out; its final verdict ultimately depends on a fine judgement of human nature."[9]

The Italian Jewish historian Sergio Luzzatto, an agnostic, in his study of the sociological setting of the Padre Pio phenomenon, writes, "All those seeking answers -- affirmative or negative -- as to whether the stigmata or the miracles were 'real' had better close this book right now."[10]

In the case of Padre Pio, the possibility of a psychosomatic origin for his stigmata is utterly implausible: he was losing one to two cups of blood every day; his wounds were the size of coins; he had no infections or anemia although he suffered the wounds for fifty years; he had no scarring where the wounds appeared; he showed no signs of neurosis and was cheerful and well-balanced unlike the so-called hysterics. Also, it is known that his wounds used to exude a fine fragrance – and those who retained relics of the bloodstains have sometimes reported that the fragrance continued over time. If someone said his stigmata were psychosomatically produced, they would have to explain how he could produce blood that did not smell like blood, a chemical marvel!

These dimensions are highlighted by one of his biographers: "For the next fifty years they [the wounds] would confound impartial science; their continuous and profuse effusion of blood, accompanied often by the sweetest fragrance, came to be regarded as a prolonged miracle, because, as the experts correctly state, blood for its production requires nourishment while this friar's extraordinary frugality was such as hardly to maintain the life of a small child. The remarkable nature of this miraculous gift becomes more apparent when it is considered how such loss of blood was simply inconsonant with and disproportionate to the stamina and energy with which P. Pio with ever greater activity and zeal conducted his life in all matters relating to the service of God."[11]

Fr. John Aurilia, an American priest, spent time with Padre Pio in his later years. "What impressed Fr. Aurilia the most about Padre Pio was his remarkable stamina, which had no natural explanation. The Padre was an old man in poor health when Fr. Aurilia worked with him. He had borne the wounds of the stigmata for 50 years. Fr. Aurilia personally witnessed how his bandages would have to be changed as many as three times a day on account of the flowing blood. Despite this, he was active from 3:00 in the morning to 11:00 at night. He prayed constantly. Fr. Aurilia would always see him with the Rosary in hand and once asked asked how many he recited each day to which Padre Pio replied, "I can't keep count."[12]

Unlike most other stigmatists, Padre Pio's stigmata were examined by medical professionals on several occasions. Although he did not wish to either display or talk about the stigmata, he allowed this examination on the orders of his superiors. The first doctor to examine the stigmata over a period of time, Dr. Luigi Romanelli, a surgeon from the town of Barletta, said, "I am certain that these wounds are not superficial because, putting my thumb in the palm of the hand, and the index finger on the back, and applying pressure on the membranes, I have the exact perception of a void existing. Those membranes are covering a hole that starts in one side, and ends in the other side. The pressure on the membranes causes intense pain."[13] He concluded, "The etiology of Padre Pio's lesions cannot possibly be of natural origins. The agent that produced them must assuredly be sought in the realm of the supernatural."[14]

As we will see, the other doctors who examined him on different occasions, were similarly convinced that the wounds were both real and inexplicable.

When asked once whether the stigmata were psychosomatically produced, Padre Pio took the bull by the horns, so to speak, and said: "Go out to the fields and look very closely at a bull. Concentrate on him with all your might. Do this and see if horns grow on your head!"[15]

And when asked if they were painful, he laughed and replied: "Do you think the Lord gave them to me for a decoration?"[16]

The very real suffering caused by the stigmata was accompanied by the hostility of the Church. As Luzzatto writes, Padre Pio's opponents "didn't hesitate to speak up soon after the events of September 20, 1918. There were some secular adversaries, obviously: skeptics and anticlericals, liberals and socialists. But also, and above all, Padre Pio had his clerical adversaries. Men of religious orders and not, modest church officials and powerful Vatican insiders."[17]

We will be examining the motives, methods and slanderous fabrications of Padre Pio's clerical enemies in a later chapter. But these are well summed up in this interview with Fr. Luigi Peroni, the author of a definitive Padre Pio biography:

> *What could explain such a persecution? And who were his persecutors?*
>
> Peroni: Since harassment of Padre Pio lasted such a long time, the motives, perpetrators and methods were many and diverse. At the beginning, it was the local Church which showed itself most hostile to the Capuchin friar. The diocesan hierarchy at different times attempted to banish him, to reduce him to silence, and to discredit him in every way.
>
> *Was the Church in Rome also opposed to Padre Pio?*
>
> Peroni: No. From Rome Padre Pio received contradictory signals. In many circles, particularly those closest to the reigning Popes, he

enjoyed esteem and respect. On the other hand, the Holy Office was less well-disposed. In my opinion, Rome was lax in disproving the absurd calumnies leveled against Padre Pio by his enemies in "faraway" Apulia (the south of Italy, where Padre Pio's monastery was located). Often accusations were credited without listening to Padre Pio's defenders.

What were the principal accusations against Padre Pio?

Peroni: From the time that Padre Pio conceived the idea for his hospital, the House for Relief of Suffering, he was criticized for his attitude towards women. With the realization of the project, that is from the 1950's onwards, he was accused of the alleged misuse of funds. In fact, that charge was a cover-up for avarice in some parts of the hierarchy, when huge contributions in money and materials came pouring in for the hospital from all parts of the world. To the great sorrow of Padre Pio, during those years he was attacked for infractions against all three of his monastic vows: poverty, chastity and obedience.

Why disobedience?

Peroni: The most diverse and ridiculous pretexts were found: for example, the length of the Masses celebrated by Padre Pio. We all know to what extent Padre Pio focused his spiritual attention on celebration of the Mass. At one point he was cautioned to "follow the common usage of good priests," that is, not to exceed 30-35 minutes.

Perhaps the greatest obstacle to Padre Pio's beatification was the so-called "Maccari dossier," a series of allegations collected by the Apostolic Visitor, Msgr. Carlo Macari, in the early 1960's. The most serious of these focused, once again, on the question of Padre Pio's "immorality." Please comment.

Peroni: Unfortunately Maccari listened to the voices raising charges against Padre Pio, from whatsoever source (village gossip, the outbursts of envious fellow monks), without giving equal hearing to the countless witnesses to his purity and saintliness.

Evil rumors were spread by certain village women, jealous of what they perceived as preferential treatment by Padre Pio of others. Those rancorous villagers told scurrilous stories to harm their "rivals," without concern for the harm done to Padre Pio. The unhappy result was a besmirching of Padre Pio's immaculate conduct.[18]

Sadly, the most effective betrayals come from within the inner circle. Think of Judas' betrayal of Jesus. In his biography of Padre Pio, the Lutheran writer Bernard Ruffin highlights the chilling strategy of Padre Pio's arch-enemy, Archbishop Maccari: "Maccari summoned all the surviving members of the faction that had attempted to undermine Padre Pio more than three decades before. De Nittis was overheard remarking to Don Domenico Palladino, who had been one of Padre Pio's most unyielding detractors, "Well, Dumi, the hour of revenge has come!'"[19]

Among the Popes, Benedict XV and Pius XI had been initially skeptical of Padre Pio and imposed restrictions on his public ministry that were later removed after lengthy investigations. Pius XII was favorable from the beginning and encouraged pilgrims to visit the Padre. John XXIII was influenced by Maccari, whom he had sent, and re-imposed some restrictions. Paul VI, who was convinced of the sanctity of Padre Pio from the Fifties, removed all restrictions. It was during the papacy of Paul VI, in 1968, that Padre Pio passed away. John Paul II was the only Pope to have met Padre Pio – which he did as a priest in 1947. In 1962, as a bishop, he wrote to Padre Pio asking for his prayers for the healing of a friend, Dr. Wanda Poltawska, who was diagnosed with cancer. Padre Pio told his secretary that this was a request he could not refuse and shortly after Dr. Poltawska was healed. John Paul visited San Giovanni Rotondo two more times – as a Cardinal in 1974 and as Pope in 1987. It was John Paul who canonized Padre Pio in 2002.

Although Padre Pio was buffeted back and forth by the ecclesiastical intrigues playing out at the Vatican, he never once resisted any orders sent by his superiors. Obedience always came first:

> Padre Pio understood it was not man one served but God. He also understood that it was through obedience to the order and leadership of man which God had put in place that he obeyed

God. Pio often taught those around him to be patient when they did not want to be and to obey when they wanted to fight. (…)

Pio gives us a modern example of how, even when we disagree or do not believe the decisions made are for the best, we should still obey. As Padre Pio worked to build a hospital to ease suffering and care for the poor, those jealous of his attention and following sought to steal the money and push Pio out of control of the leadership of the hospital. Pio remained silent and obeyed. He remained focused on the main thing, which was his personal obedience to God and others. The attention gained by the stigmata and miracles reported under Pio caused the Church leadership, even some of his own religious brothers, to lie on him and put him through unnecessary investigations. Pio still did not defend himself or fight back. He was obedient. As a result, he was justified in the end, even to the point where his leadership of the hospital was restored and he even received dispensation from his vow of poverty by Pope Pius XII in order to enable him to oversee expenses of the hospital. Pio, in another sign of obedience, gave the hospital to the Vatican.

False accusations by his own religious brothers led Pio to be ordered to stop celebrating Mass publicly and he was to celebrate it only privately in his room. He obeyed. He was separated from his spiritual advisor and was to never contact his advisor again. He obeyed. When he was allowed to finally say Mass publicly again, it was only to be at a 4 a.m. Mass so the leadership could reduce the size of the crowds gathering there (or at least they hoped it would, but it failed). Padre Pio obeyed without questioning.

Many have heard of St. Padre Pio's battles with the devil, suffering of the stigmata, and perhaps even his work with the hospital. However, it is his greatest legacy of simple obedience that makes this friar more relevant for our time than ever before. This simple, poor friar who offered himself for others and loved others more than himself, forgave those who attacked him and lied on him even if they never asked, and sought only to do the will of God, left us the greatest lesson in the way to holiness. Obedience.[20]

His personal response when the Vatican stripped him of the faculties of his priestly ministry was moving and exemplary:

> The superior of the monastery read to Padre Pio the decree and the saintly friar responded, "God's will be done ... The will of the superiors is the will of God."

> Padre Pio would spend the next several years in silence, celebrating Mass privately and accepting no visitors.

> He couldn't even write to his spiritual children.

> During this time many came to Padre Pio's defense, submitting petitions, writing letters and even publishing books.

> Padre Pio was appalled by this response and urged them to stop, writing to the local bishop to assert his dissociation with them.

> "I must repeat that I am very disgusted by the unworthy behavior of certain false prophets who speak in my behalf ... they should stop this false and unworthy propaganda, but meanwhile they have followed in their morbid fanaticism, not caring about the supreme authority of the church.

> I turn, therefore, as a son most humble and completely obedient to the Catholic Church ... With profound humility I kiss your sacred ring and profess myself to your excellency your most humble and obedient son."

> Above all, even though the accusations put against him proved false (and he would eventually be allowed to minister without restrictions), Padre Pio submitted to the decree and remained silent, trusting in God's provident plan.[21]

What did Padre Pio think of Benito Mussolini, the Fascist ruler of Italy, and of Italy's role in the Second World War? Ruffin writes,

> Padre Pio feared and despised Adolf Hitler because of his "religion of blood" (...) One day, after dinner, when the friars were discussing the situation in North Africa, where Italian and

German troops were battling the Allies, Padre Pio commented, "We lack the [military] means to win, and also the help of God, because Mussolini, and, especially, Hitler, are too evil." He told another priest, deploring Italy's alliance with Germany: "A people chosen by God uniting with the enemy of God – this is something that will lead to punishment! (…) We will retreat to Bengazi and then into the desert, then to Tripoli and then into Sicily and then into Italy." (…) His prediction was proven correct."

He furthermore maintained that the Lord will never allow a man as diabolical as Hitler to win. (…)

When the Allies invaded Italy in 1943, a desperate Mussolini sent messengers to Padre Pio, requesting his prayers for Italy. "So, *now* you come to me, after you have destroyed Italy!" he fumed. "You can tell Mussolini that nothing can save Italy now! Nothing! You have destroyed her!"[22]

Man with a Plan

The stigmata were just one of the supernatural elements of the Padre Pio phenomenon. It was the most prominent because it was the most visible.

But Padre Pio was well-known for multiple other extraordinary charisms. Like the stigmata, these charisms were given to him by his Savior for one purpose – to assist in the mission of salvation and sanctification by drawing multitudes back from the brink of eternal ruin.

He was not bound by the normal constraints of space. He was literally all over the map. Bilocation came naturally to him. This is the capability of being in two places at once. Although he never physically left San Giovanni Rotondo he is known to have appeared in many other places – from Hungary to Czechoslovakia, from America to the Pacific Ocean, and, of course, in various cities in Italy. Wherever God wanted to send him. There was always a purpose for his "flying visits" – to help someone who needed help.

He was also partially free of the limitations of time. His time zones extended to the future! Many people have occasionally had precognitive experiences where they "see" something that takes place in the future. With Padre Pio, this again was something that seemed to almost come "naturally." He was able to warn people who were in imminent danger or who needed spiritually urgent advice. This charism of "seeing" the future, like his bilocation, was given to him to serve his fellows.

Additionally, Padre Pio peered into other dimensions of reality that are inaccessible to us in our current state of being. Of these, the one he was most concerned with was the realm of those who had died in the friendship of God but needed to be cleansed of earthly flaws before entering the presence of God. Such cleansing is a painful process. This is the state of being often called Purgatory (for "purge" or cleanse). It was yet another avenue for him to minister to those in need. The souls in Purgatory need our prayers since they cannot help themselves. Padre Pio was often engaged in conversations with the souls who came for his help or thanked him for helping them with his prayers and Masses.

Inevitably, Padre Pio's work for God attracted the attention of the arch-enemy of all that is good, the Devil. The Padre's ministry was drawing sinners away from the captivity of the Evil One. So the Devil retaliated at two levels. No good deed would go unpunished. First, he physically attacked Padre Pio often. In addition to the wounds of the stigmata, the Padre suffered bruises and gashes inflicted by this dark adversary. Secondly, the Devil was called "the father of lies" by Jesus because he is the source of deceit and division. This was the other mode of diabolic attack. Padre Pio was constantly subjected to outrageous slanderous attacks that were works of deceit and division.

Healing people through the Power of God was another task entrusted to the Padre. The healing miracles attributed to the prayers of Padre Pio include the restoration of sight to a girl born without pupils. These miracles came from the Hand of God and the Padre was simply an intercessor. But his intercession bore powerful results as even the future Pope from Poland discovered. Yet, Padre Pio did not simply rely on miraculous interventions but was also committed to making the natural means of healing available to all. To this end, he spent considerable time and energy in building a hospital that is now one of the largest in Italy.

No sinful secret was safe with those who went for confession to Padre Pio. He would sometimes – accurately – list the sins of the penitents if they were hesitant or forgetful. He even "knew" the thoughts of some of his spiritual children when they were hundreds of miles away. Yet again, this charism was exercised purely to draw the recipient of the miracle closer to God.

Padre Pio made his presence felt through the sense of smell. Fragrance was his calling card! His spiritual children knew he was with them when they smelt roses or some fine fragrance "out of the blue." His wounds, we have seen, had the smell of perfume.

Padre Pio was always conscious of his responsibility for the eternal well-being of all those who came to him – whether physically in person or mentally by intention. So, almost from the beginning of his ministry, he became a spiritual father to all who sought his help (much like St. Paul was the spiritual father of the Corinthians and the Thessalonians and all others to whom he preached the Gospel). Even today, those who wish to be his children can "sign up" at the monastery of San Giovanni Rotondo. Padre Pio's spiritual fatherhood had a unique dimension: he promised that he would not enter Heaven until all his spiritual children were ushered in. So great was his love.

Padre Pio lived in many dimensions. But he was as normal as they come: humorous, moved to anger when this was called for, kind and compassionate. Moreover, his gifts were not "powers" he could exercise on demand. He had to pray for divine assistance anytime someone came to him with a need and not every such request would result in the desired answer. He reminded those who came to him that all he could do was ask – the response was entirely dependent on God.

Padre Pio did not see himself as a miracle-worker. Nor did he want the focus to be on him since he was simply an instrument of the divine. He sought only one thing – to turn people to God.

Padre Pio, writes Ruffin,

> told an American Red Cross official: "People think I have miracles. I have no miracles." When people thanked him for a supernatural

favor, he usually told them, "If you think you have received a grace, go to Our Lord and thank him, not me."[23] (…)

One day Padre Pio showed another priest a letter addressed to him and read the salutation aloud: "To the sainted Padre Pio (…) *Sainted!* (…) *How beautiful!* (…) To the *sainted* Padre Pio!" "You're not getting proud, are you?" asked the other priest. Spreading his arms and lowering them to imitate someone boasting, Padre Pio quipped, "Oh, yes, my friend, I'm getting quite proud of it now!" and burst out laughing. For him, being considered a saint was a joke.[24]

Ghezzi reminds us of his "normalcy":

Focusing too much on Padre Pio's marvels and mystical phenomena gives the false impression that he led an abnormal life, more angelic than human. While he opened our eyes to heavenly realities, he kept his feet firmly planted on the earth, enduring and enjoying ordinary things, as other human beings did. Today we mainly imagine him as a wonder-working stigmatic with miracles flowing from his wounded hands. But the people who knew him, while they appreciated his marvels, loved him more for his earthiness, his compassion, his gentleness, his humor, and his common sense. For instance, when he was asked his opinion of a thief who had stolen valuable gems from a church's painting of the Virgin, he responded, "What do you want me to say? That poor young man was probably hungry and went to Our Lady to say: 'Of what use are these jewels to you?' And probably Our Lady gave them to him. Silly him to get caught with the goods in his pocket."[25]

Ruffin adds,

It would be a serious mistake to assume that life with Padre Pio was a continual experience of the supernatural and the paranormal. Many of those who lived with him never saw or experienced anything unusual. The priests and brothers who knew him well knew him as a normal human being whose distinguishing characteristics were kindness, joy, serenity, and humility. (…) Padre Marcellino Iasenzaniro wrote, "The Padre was severe and sharp only in two cases: in the confessional, when striking and denouncing sin as evil, the real evil that harms man; and when one showed excessive veneration for him as a cult figure. (...) In all other circumstances (...) he was the most humble, exquisite, gentle, kind, caring, and affectionate person I have ever met." [26]

Padre Pio was a prankster and also had a ready sense of humor with which he regaled his fellow monks and spiritual children. As can be imagined, he was constantly receiving entreaties for his intercessory prayer. Sometimes the requests were outlandish. Ruffin mentions one such instance:

One girl sent a request plastered with clippings. "I am a girl, unfortunately not pretty enough," she wrote to Padre Pio, "And I come to ask your prayers to help me." She informed him that she wanted a nose like that of a particular actress, enclosing a picture of the woman she wanted to resemble. She went on to say she wanted lips like another actress and the ears and eyes of a certain model, and enclosed likenesses of both. After Padre Eusebio translated the letter for him, Padre Pio, asked, "Did she say anything about a brain?" "No, Padre she didn't say anything." "Then answer that I'm praying that she might get a brain."[27]

But he was the sinners' saint: always aware of the depths of human depravity but serving as a beacon of the life to which we are called as Ruffin points out:

In 1962, a woman wrote him from London, "Padre, I'm a prostitute. Every evening, at nine, I'm dragged out into the street. I am ashamed. I am writing to you so that you might help me overcome this shame." She didn't ask to be freed from her "protector," or her profession, only from her shame. Padre Eusebio, who expected Padre Pio to give a severe warning to repent from her sin, was astounded when he said, "Answer that I will pray for her with all my heart." After a while the woman wrote back, "Dear Padre, thanks to God after receiving your letter, nine o'clock comes, but I no longer go down on the street. I succeeded in ridding myself of the one who dragged me here." "*Deo gratias*," he responded. [28]

The famous novelist Graham Greene, a great chronicler of the darkness of human nature, was understandably impressed and disturbed by the power of Padre Pio:

> I have also seen with my own eyes the stigmata on Padre Pio's hands in the south of Italy. He was not a man who looked as though he suffered from a nervous disorder. I was so convinced of his powers of goodness that I refused to approach him and speak with him. I explained to the friends who had brought me along that I was too afraid that it might upset my entire life.[29]

The Greatest is Love

In all that he did, Padre Pio wanted to become more like Jesus, the Man of Sorrows. Jesus asked us to take up the cross and follow him. Jesus said that there is no greater love than that we give up our life for our friends. Padre Pio took him at his word.

He also identified his constant pain and suffering with the blueprint laid out by St. Paul: "I rejoice in my sufferings for your sake, and in my flesh I am filling up what is lacking in the afflictions of Christ on behalf of his body, which is the church." (*Colossians* 1:24)

But Padre Pio did not see his life of suffering as one to which all are called. Says Ruffin:

> Although Padre Pio embraced a ministry of suffering, he usually discouraged people from seeking suffering beyond that which they experienced in the course of life. It was enough to endure that in patience and good cheer. (…) When another woman, in order to do penance for her sins, ate poisonous weeds, Padre Pio sternly rebuked her, "I won't permit any more of this madness! There are many ways to do penance, such as offering to the Lord whatever trouble comes to us day by day. It's up to the Lord to give us our cross. If he hasn't sent you one, it's because he's not sure that you could bear it."[30]

Padre Pio always wanted to give of himself to others – and encouraged them to keep asking for more! The Englishman Cecil Humphrey-Smith, a recipient of his miracles and one of the Padre's frequent visitors, writes that "when I apologized for calling on him so often, he laughed at me a little and told me, 'Be generous, ask for more.'"[31]

Padre Pio passed away on September 23, 1968, at the age of 81. As we have seen, his desire was to take on "the punishments prepared for sinners" until the end of the world so as to bring about their conversion. He said too, "I belong entirely to everyone" and "Everyone can say: Padre Pio is mine."[32]

The Padre Pio phenomenon rests on the solid foundation of hard facts like his naturally inexplicable stigmata; the testimonies of those who witnessed multifarious miracles (from bilocation to miraculous healings) while he was alive; and the continuing series of encounters with Padre Pio enjoyed today by those who become his spiritual children.

The Padre Pio phenomenon rests also on the legacy he left of continual suffering offered up for his fellows, relentless shaping of souls at the confessional and spiritual fatherhood to the thousands who turned and turn to him.

Padre Pio was truly Jesus' gift to the modern world.

In the sections that follow, we will explore in more detail the charisms of Padre Pio, within the framework of scientific paradigms and rational thought, and conclude with the endgame to which they were directed – the Promised Land beyond space and time.[33]

Notes

[1]Excerpted from *Mystics and Miracles by Bert Ghezzi*, https://www.loyolapress.com/catholic-resources/saints/saints-stories-for-all-ages/blessed-padre-pio-of-pietrelcina-1887-1968/
[2]*Gente*, June 27, 2002, 21–25.
[3]https://caccioppoli.com/Close%20encounters%20with%20Padre%20Pio%20in%20the%20Confessional.%20Baptisms,%20Communions,%20Weddings.html
[4]https://wwwn.cdc.gov/TSP/MMG/MMGDetails.aspx?mmgid=144&toxid=27

[5]https://www.catholicleague.org/60-minutes-attacks-padre-pio-and-mother-teresa/

[6]https://www.ncbi.nlm.nih.gov/pmc/articles/PMC3854941/

[7]John Hardon, *Catholic Dictionary* (New York: Doubleday, 1985), 420.

[8]https://www.livescience.com/42822-stigmata.html

[9]https://www.smithsonianmag.com/history/the-mystery-of-the-five-wounds-361799/

[9]Sergio Luzzatto, *Padre Pio: Miracles and Politics in a Secular Age* (New York: Metropolitan Books, 2007), 5.

[11]Augustine Mc Gregor, "The Spirituality of Padre Pio," https://www.ewtn.com/catholicism/library/stigmata-of-padre-pio-13854

[12]https://www.catholicworldreport.com/2021/11/02/a-secretary-of-padre-pio-remembers-the-indefatigable-saint/

[13]https://caccioppoli.com/6%20Examinations%20of%20the%20wounds.html

[14]Luzzatto, op cit., 37.

[15]C. Bernard Ruffin, *Padre Pio: The True Story* (Huntington, IN: Our Sunday Visitor, 2018), 412.

[16]*Who is Padre Pio?* (Rockford, Illinois: Tan Books, 1974), 9.

[17]Luzzatto, op cit., 6-7.

[18]Andrea Monda, "And the Light Shone in the Darkness," *Inside the Vatican*, April 1999, 8-9.

[19]Ruffin, op cit., 283-5.

[20]https://www.catholic365.com/article/11364/st-padre-pio-modern-saint-of-obedience.html

[21]https://aleteia.org/2021/07/13/how-st-padre-pio-responded-when-the-vatican-silenced-him/

[22]C. Bernard Ruffin, op cit., 283-285.

[23]Ibid., 279.

[24]Ibid., 271.

[25]Ghezzi, op cit.

[26]Ruffin, op cit., 271.

[27]Ibid., 429.

[28]Ibid., 429-30.

[29]"The Uneasy Catholicism of Graham Greene," *The New York Times*, April 3, 1983.

[30]Ruffin, op cit., 343.

[31]"A Saint on My Back," *The Voice of Padre Pio*, November/December, 2007, 20.

[32]https://caccioppoli.com/Padre%20Pio%20in%20his%20own%20 words%20about%20hymself,%20God,%20Jesus,%20Holy%20Spirit,%20 Church.html

[33]Many of the citations listed here are taken from Caccioppoli.com a comprehensive, reliable website on Padre Pio that was created by one of his beloved spiritual children, Dr. Giuseppe Caccioppoli.

Quantum "Splitting" –
In Two Places at Once!

Each one of Padre Pio's reputed prodigies was mystifying in its own right. But accounts of his bilocation were especially baffling. Bilocation, as the word suggests, is the ability for one person to be in two locations at the same time. Historically a few saints were reported to bilocate. The famous bilocators included St Anthony of Padua, St. Martin de Porres and others.

Although Padre Pio spent almost his entire life in one location, a remote monastery in the south of Italy, he somehow managed to be all over the world as and when required!

Consider these accounts.

The Flying Monk

> In the midst of World War II, Italy was invaded by Nazi Germany and Allied forces made many attempts to liberate the country. According to various accounts, intelligence reported a cache of German munitions near San Giovanni Rotondo, the town in which stood the monastery of St. Padre Pio.
>
> However, at the beginning of the war Padre Pio reassured the people that no bomb would touch their small city. True to his word, Padre Pio reportedly went out of his way to make this happen.
>
> According to author Frank Rega in his book *Padre Pio and America*, "none of the Allied planes sent to bomb the San Giovanni

Rotondo area were able to complete their missions successfully. There were often mysterious malfunctions, causing the bombs to drop harmlessly in the fields, or mechanical failures which caused the planes to veer off course."

Most remarkable of all were the stories of a "flying monk."

An American pilot was just about to bomb the city when, "Suddenly, the pilot saw in front of his plane the image of a monk in the sky, gesturing with his arms and hands for the plane to turn back. The shocked pilot did just that, and jettisoned his bombs elsewhere. When he returned to the base and told his story, his commanding officer decided it was best to put this pilot in a hospital under observation for mission-fatigue."

The pilot couldn't get the image from his mind and after the war he made inquiries to find this monk. He eventually made the journey to San Giovanni Rotondo and recognized the "flying monk" as St. Padre Pio.

While the accounts might appear to be fictitious tales, what is true is that San Giovanni Rotondo was spared during World War II and multiple pilots reported seeing a "flying monk" in the sky over the spot where they were ordered to unload bombs.[1]

Pacific Ocean

In 1946, an American family went from Philadelphia to San Giovanni Rotondo in order to thank Padre Pio. In fact, their son, a bombardier plane pilot (during World War II), had been saved by Padre Pio in the sky over the Pacific Ocean. The son explained; "the airplane was flying near the airport on the island where it was going to land after it had loaded its bombs. However, the airplane was struck by a Japanese attack plane. The aircraft exploded before the rest of the crew had the chance to parachute. Only I succeeded in going out of the airplane. I don't know how I did it. I tried to open the parachute, but I didn't succeed. I would have smashed to the ground if I had not received a friar's help who had appeared in midair. He had a white beard. He took me in his arms and put me sweetly at the entrance of the base. You can imagine the astonishment inspired by my story. Nobody could believe it, but given my presence there, they had no choice. I recognized the friar who saved my life some days later while on home leave, I saw the monk in one of my mother's pictures. She told me she had asked Padre Pio to look after me."[2]

Historically, there was always a good reason for a person's bilocation. It provided a way for the bilocator in question to perform an urgent spiritual service in a place geographically separate from the location of their body. In daily life, when people are confronted with urgent tasks to be performed in multiple locations, they sometimes exclaim, "I can't be in two places at once." When confronted with this kind of a problem in the world of salvation history, God enables the bilocator to be in "two places at once" as required.

Does bilocation mean that the physical body of the bilocator is in two places at once? No. This cannot be the case because to be physical means to occupy a specific "block" of space and to operate "from" there. Thus, the exact same set of quantum fields that make up a body cannot simultaneously occupy two geographically separated "blocks" of space. What happens in bilocation is that the *person* to whom the body belongs is able to be present in a second location. This is possible because a human person cannot be reduced to a "physical" entity but is an agent who operates the unique union of spirit and matter that we call a human being.

How do we know that the human person transcends the physical? Although we can associate specific brain events with various mental acts, these correlations do not include the most fundamental mental reality – *me*, the "I" who is conscious and does the thinking. The "I" is not present in any region of the brain.

But how do we know the self, the "I", exists? What is most obvious in my everyday experience is the fact that I do experience things – and that this "I" which does the experiencing exists and that its experiences are "my" experiences.

The existence of the self is especially apparent at a biological level. Our brain is a flux of chemical reactions, electrical impulses and constant molecular change. We could not even complete a sentence or remember what happened a minute ago if there was no "I", the soul, that is stable, enduring and non-physical.

In his book, *How the Mind Works*, MIT professor Steven Pinker, writes, "The 'I' is not a combination of body parts or brain states or bits of information, but a unity of selfness over time, a single locus that is nowhere in particular."[3]

Not being a physical entity, the soul of a person is in theory not limited by matter and can be "present" in a location distant from the body. This is what happens at death. But here on earth, human souls operate only through their bodies unless divinely enabled to transport themselves "from" their bodies to a geographically different location.

And this is what happens in bilocation. Thus the bilocator's body stays, say, in Moscow while the bilocator is "present" in Bogota. The presence of the bilocator is identified in Bogota because someone recognizes what seems to be their body and its "operator", namely the bilocator. The second "body" is not physical: it is an icon much like the appearances of angels in the Bible and history. The bilocator is in Bogota to perform something important in the plan of God.

Some who reject various accounts of bilocation tell us that science shuts the door to its possibility. But the quantum physics that is a fundamental element of modern science makes the claim that a particle can be at two places at once. What this means is that the effect of a single quantum field can be measured in two different locations. This is what we are claiming with bilocation. Once you understand both the non-physical nature of the human soul and its ability to bring about effects in the physical world, you implicitly grant the possibility that it can potentially exercise an effect in more than one physical location.

Although bilocation is a profound mystery, this particular Padre Pio prodigy was not unique to him. As we have noted, there have been other bilocators in history.

One well-known bilocator who is reported to have visited the United States from Spain in the 17th century without leaving Spain is Mary of Agreda.

Mary of Agreda (1602-1665) was a Spanish nun who has been declared "venerated" (a step in the process of becoming a saint). She wrote about mystical experiences and became known for her own experience with them through bilocation.

Even though Mary was cloistered inside a monastery in Spain, she reportedly appeared on various occasions to people in Spanish colonies in the area that would become the United States of America. Angels helped transport her to the New World from 1620 to 1631, she said, so she could speak directly to Native Americans from the Jumano tribe living in what is now New Mexico and Texas, sharing the Gospel message of Jesus Christ with them. Angels translated her conversations with members of the Jumano tribe, Mary said, so even though she spoke only Spanish and they spoke only their tribal language, they could still understand each other's language.

Some of the Jumano people contacted priests in the area, saying that a lady dressed in blue had urged them to ask the priests questions about faith. Mary always dressed in blue, since that was the color of her religious order's cape. A variety of church officials (including the Archbishop of Mexico) investigated reports of Mary bilocating to the New World colonies on more than 500 separate occasions over 11 years. They concluded that there was ample evidence that she had actually bilocated.

Mary wrote that God has given everyone the ability to develop and use spiritual gifts. "So great is the impetus of the river of God's goodness overflowing on mankind ... if creatures would place no obstacle and permit its operations, the whole soul would be inundated and satiated with participating in its divine essence and attributes," she wrote in her book The Mystical City of God.[4]

A native of San Angelo, Texas, the location to which she is supposed to have been transported, reports that Mary of Agreda is today a familiar name in his city:

> During her many bi-locations to West Texas she encountered Jumano Native Indians to whom she preached and who were later converted to Christianity. And while the nuns in the monastery testified that Sor Maria never left Agreda, the Jumanos could describe her in detail and referred to her as the Lady in Blue because of the blue cape she wore all the time.

> According to the Jumanos, as a sign of her final appearance, the Lady in Blue left a meadow of beautiful blue flowers. It is now the state flower of Texas – the bluebonnet.

> Nine years ago, the Most Rev. Michael D. Pfeifer, then bishop of the Diocese of San Angelo, proclaimed June 20 as "Lady in Blue Day" for the City of San Angelo, and for the entire diocese consisting of 26 counties in West Texas, a befitting tribute to the Venerable Sor Maria de Jesus de Agreda, currently being considered for canonization as a saint.[5]

What was distinctive about Padre's Pio's bilocation was its frequency, the many who witnessed it and the fact that it took place in the modern age. Patricia Treece's book *The Sanctified Body* described his gift of bilocation. "Except for being driven to the polls to vote, from 1918 to his death in 1968, Padre Pio never left the environs of the Friary of Our Lady of Grace. Yet there are many testimonies, as one Capuchin who knew Pio puts it, that during that same half-century, 'Padre Pio was popping up all over the place.'"[6]

Here are a few reports of Padre Pio's "travels" taken from sources connected to his monastery.

Padre Pio at the Beatification and Canonization of St. Therese of Lisieux

> Saint Therese of Lisieux, whose feast day is October 1st, had a great influence on Padre Pio. (...)

Although physically present at the friary in San Giovanni Rotondo in Southern Italy, Padre Pio was seen at the beatification of Saint Therese in the basilica of Saint Peter in Rome on the morning of April 29, 1923. He remained in the basilica until a prelate tried to approach him to speak to him, at which time he vanished. This incident was confirmed by the Archbishop of Salto, Uruguay, Msgr. Thomas Gregorio Camacho.

The second incident is even more wonderful because of the authority of the witness who saw him: Blessed Don Luigi Orione, founder of the Little Work of Divine Providence. Only two years after the beatification and twenty-eight years since her death, the saintly Pontiff Pius XI canonized Saint Therese on May 17, 1925. The ceremony was again held in the basilica of Saint Peter, but this time it was so full that most of the devoted had to remain outside in the square. (…)

Padre Pio was present in the midst of all this (…). Yet, in fact, he had not moved from San Giovanni Rotondo. He was seen clearly by Don Orione, just as he had been seen on another occasion at the Vatican praying at the tomb of Saint Pius X, to whom he had a special devotion. "But," Don Orione recounted (…), "whenever I approached him, Padre Pio would vanish."[7]

Padre Pio Visits Cardinal Mindszenty at his Jail Cell in Hungary

Andrea Tornielli relates that [Padre Pio: la sua chiesa, I suoi luoghi, tra devozione, storia e opera d'arte" ("Padre Pio, His Church and His Places: Devotion, History, and Works of Art") by Stefano Campanella] contains the story of Angelo Battisti, director of the Casa Alivio del Sufrimiento (House for the Relief of Suffering), and typist for the Vatican Secretariat of State. Battisti was one of the witnesses in the process of beatification of the holy friar.

Cardinal József Mindszenty, archbishop of Esztergom, Prince Primate of Hungary, was imprisoned by the Communist authorities in December of 1948 and condemned to life in prison the following year.

He was falsely accused of conspiring against the Socialist government. He was in prison for eight years, then under house arrest, until being freed during the popular uprising of 1956. He took refuge in the United States' Embassy in Budapest until 1973, the year in which Paul VI forced him to leave and to renounce his archdiocese.

During those years of prison, Padre Pio appeared in Cardinal Mindszenty's prison cell by bilocation.

In the book, Battisti describes the miraculous scene as follows:

"While he was at Saint Giovanni Rotondo, the Capuchin who bore the stigmata went to bring the Cardinal bread and wine destined to be transformed into the body and blood of Christ..."

"As is known," Battisti explains, "Cardinal Mindszenty was taken prisoner, thrown in prison and kept in sight by the guards at all times. As time passed, his desire to be able to celebrate Mass became very intense."

"A priest who came from Budapest spoke to me confidentially about the event, asking if I could obtain confirmation from Padre Pio. I told him that if I had asked about something like that, Padre Pio would have scolded me right out the door."

But one night during March of 1965, at the end of a conversation, Battisti asked the stigmatized Capuchin:

"Father, did Cardinal Mindszenty recognize you?"

After a first irritated reaction, the saint answered, "We met and we had a conversation, and you think he might not have recognized me?"

This confirmed his bilocation to the prison, which had taken place several years earlier.

"Then," Battisti added, "Padre Pio became sad and added, 'The devil is ugly, but they had left him uglier than the devil,'" referring to the maltreatment that the Cardinal suffered.

This shows that Padre Pio had brought him help from the beginning of his time in prison, because humanly speaking one cannot conceive of how the Cardinal was able to resist all the suffering to which he was subject, and which he describes in his memoires.

Padre Pio then concluded, "Remember to pray for that great Confessor of the Faith, who suffered so much for the Church."[8]

Padre Pio in New York

Ellie Hunt's family came from Padre Pio's hometown of Pietrelcina and had known Padre Pio from his childhood. They eventually emigrated from Italy to New York. In 1960, when Ellie was 31 years old her grandfather, Jack Crafa became gravely ill. Ellie and her parents lived close to his home in Flushing, New York and during his grave illness, the family stayed by his side. When Jack eventually fell into a coma, they all knew that his life was soon passing.

One day while Ellie and her parents were at her grandfather's bedside, a stranger knocked at the door. They were all surprised to see a Capuchin monk dressed in a dark brown habit, because there were no Capuchin monks in their Parish or in any other parish in the area for that matter. They were also surprised to see that he was wearing sandals without any socks, because it was a particularly cold day and snow covered the ground. The monk said that he had come to pray for her grandfather. Ellie was also a bit upset, because she thought that the parish priest should have come to pray for her grandfather, and not a complete stranger. However, she was soon impressed by the kindness and compassion of the young monk.

He then went straight into her grandfathers bedroom and blessed Jack Crafa. Then told the family to pray the Rosary suggesting that they sit at Jack's side, praying the Hail Mary close to his ear, for he seemed to have the opinion that Jack was still able to hear. After the monk said that, Ellie was surprised to find that when she took her grandfather's hand in hers, she felt a response from a very slight squeeze from his hand.

The young Capuchin then gave Ellie's grandfather the Last Sacraments, then he blessed the family and bid them goodbye. As he walked out the front door, Ellie's father, James noticed that there was no car waiting for him outside. James watched him as he walked up the street until he disappeared in the darkness. Jack Crafa died that very night. He had been in a coma for nine days.

After the unknown monk had left, Ellie's father James became pale and appeared quite shaken. Ellie's mother Lucy, asked him for the reason. "Don't you know who that was?" James replied, "It was Padre Pio. He came to give the Last Rites to your father and he looked exactly like I remember him when I used to deliver eggs to him in Pietrelcina."

For Ellie's grandfather Jack had been one of Padre Pio's spiritual sons from Pietrelcina and through a special grace God Padre Pio had come to comfort and encourage him, and to administer the last Sacraments of the Church, for his death was only hours away.[9]

Fr. Carty's Reports

Fr. Charles Carty from the US visited Padre Pio several times in the 1950s and wrote the first book on Padre Pio in English. Here are some instances of Padre Pio's bilocation brought to his attention.

Many saints experienced only the first phases of this supernatural gift, that is of being raised from the ground, which is known as levitation, whereas in the case of Padre Pio, the bilocation was complete. He moved his body about from one place to the other.

One of his forms of bilocation was through the use of his voice. This happened in the case of many of his spiritual children as well as to strangers at any distance from him, bringing news, or giving advice or reproaching them as he so often did in dreams. In this last case, the fortunate one heard his voice clearly and distinctly without seeing him.

Telegrams, telephone transcripts, testifying letters, and many contacted eyewitnesses gave testimony of the foregoing bilocations and that Padre Pio was seen throughout Italy, in Austria, distant Uruguay, and Milwaukee, Wisconsin.

Visit to Uruguay

Msgr. Damiani, Vicar General of the diocese of Salto, Uruguay, South America, was a faithful friend of Padre Pio and came to visit him repeatedly. Upon one such visit in 1929 he said to Padre Pio: "I would like to die here so you could assist me." Padre Pio answered: "No, you will die in Uruguay." "Will you come down there to assist me?" Padre Pio said, "Yes."

One forenoon during this visit, the Monsignor had a slight heart attack. He sent for Padre Pio immediately. But, since he was hearing Confessions, he did not heed the call. When he came upstairs about noon and went to the room of Msgr. Damiani, the latter chided him in a friendly way: "Padre Pio, why did you not come when I sent for you? I could have died." Padre Pio answered with a smile: "Man of little faith. Did I not tell you that you would die in Uruguay?"

During Holy Week of 1949, Most Rev. Antonio Maria Barbieri, O.F.M. Cap., Archbishop of Montevideo, Uruguay, came with 38 of his diocesan people to pay Padre Pio a visit. When they arrived he asked to see Padre Pio alone first. He went to Confession to him and then had a little chat with him. Later, after Padre Pio had gone to the confessional, the Archbishop spoke with the rest of the fathers and related the following:

"In 1942, Bishop Alfredo Viola of Salto celebrated his Silver Sacerdotal Jubilee and at the same time had the laying of the cornerstone of his minor seminary. The Apostolic Delegate and five Bishops lodged in the episcopal residence. On the vigil of the Jubilee, about midnight, I was awakened by a knocking at my door. The door was open about a foot. I saw a Capuchin pass by and heard a voice: 'Go to the room of Msgr. Damiani; he is dying.' I arose, put on my cassock, called the other Bishops and some priests, and we went to the room of the Monsignor. On his night table I found a slip of paper written by Msgr. Damiani: 'Padre Pio came.' (The Archbishop had this paper with him.) Now today, when I spoke to Padre Pio after Confession a little while ago, I asked him: 'Padre Pio, were you the Capuchin whom I saw in the residence of the Bishop of Salto, the night Msgr. Damiani died?' Padre Pio was embarrassed and did not answer, although he easily could have said no. When I insisted and he still would not answer, I laughed and said: 'I understand.' Padre Pio nodded and answered: 'Yes, you understood.'"

Here we have: first, the promise of Padre Pio to assist the Monsignor at his deathbed, then the Monsignor's written testimony that Padre Pio had come; the testimony of the Archbishop and finally the practical admission of Padre Pio.

Hawaii

One of his secretaries writes that a lady wrote from Waiakoa, Kula, Mani—one of the Hawaiian Islands. She thanked Padre Pio for visiting the prison on the island of Oahu, where her husband was an inmate. Her husband had phoned her, gave her Padre Pio's address, and asked her to write. He did not go into details as to what Padre Pio had said or done, but whatever it was, the prisoners were all consoled by the visit. This was clearly another case of bilocation.

Wisconsin

Padre Pio himself admitted, when asked, that he was in Milwaukee, Wisconsin, June 25, 1950, to assist at the death of the father of a Capuchin.

Florence

On July 20, 1921, a Monsignor D'Indico of Florence, whom this author met in 1923 when studying theology at the Archbishop's Seminary at Florence, was alone in his study. He felt the sensation of having someone at his back. He turned and saw a monk, who disappeared. He left his room to tell a chaplain what happened. The chaplain thought it was mere hallucination due to his state of anxiety over his sister, who was very ill. He invited him to take a short walk for mental distraction. When they returned they called at the sick room. His sister, who a little before had been in the state of coma, at the

same hour when her brother felt the sensation of being in the presence of Padre Pio, told how she had seen a monk enter her room who approached her and said:

"Don't be afraid; tomorrow your fever will disappear, and after a few days there will be no trace of your illness on your body." "But, Padre," she answered, "are you then a saint?" "No, I am a creature who serves the Lord through His mercies."

"Let me kiss your habit, Padre."

"Kiss the sign of the Passion," and he showed his hands transfixed and bleeding.

"Padre, I recommend to you my husband and child."

"Pray, pray that you will be good, and be assured that your child will be under my protection," and blessing her, he vanished.

She immediately got better and in eight days was entirely cured.

The General

One day during the war General Cadorna was in his study absorbed in thought, and he held his head in his hands, thinking of all the young men who, for the love of country, would have to give up their lives; when suddenly he smelled a very strong perfume of roses which was wafted around the room. Raising his venerable head, he was stupefied to see a monk with a seraphic look and with bleeding hands. Passing in front of him, the monk said, "Be calm, they will not do anything harmful to you."

With the disappearance of the monk, the general no longer smelled the perfume. He told a Franciscan about the vision, and when he mentioned the perfume the Franciscan said, "Your excellency, you have seen Padre Pio." Then he told the general all about Padre Pio. The general decided to visit San Giovanni; and when he arrived there incognito, he was immediately approached by two Capuchins who had recognized the general even though he was trying to disguise himself in civilian clothing. They approached him and said, "Your excellency, Padre Pio is waiting for you. He sent us to meet you."

Prayer Group

At San Martino in Pensilis, a group of devout followers of Padre Pio and members of the Third Order of Saint Francis were in the habit of gathering in the house of one or the other for their Franciscan reunion. Before starting they would all say a little prayer to the Guardian Angels, asking the celestial messengers to go and call Padre Pio, asking him to be present in their midst. Very often, after having prayed to their Guardian Angels, they all noticed the perfume of Padre Pio, and they all reverently knelt down saying: "Padre Pio is with us; he will direct our meeting."

On one of these occasions, the meeting was in the house of the police marshal Trombetta. Giovannino, his five-year-old son, was present. Suddenly the little boy ran to his mother's knee and hid his face, saying: "Mamma, I am afraid, because Padre Pio is here. I see him." "Where, Giovannino, where? Is he still here?" they all asked. The child clung to his mother, as he peeked out and again hid his face, pointing with his finger to the place where he saw the Padre, saying, "Now he is here, now he is there. He is still with us," and finally, "He is going away now." All eyes followed the direction of the little pointed finger but saw nothing.

Several of those present went to San Giovanni for the purpose of asking Padre Pio if he had really been with them. They told him what had happened, and then said: "Padre, was it really you that Giovannino saw?" "And who else could it have been?" was his abrupt reply. He was always very embarrassed and almost painfully timid when people openly spoke to him about his supernatural faculties, but "fools jump in where angels fear to tread." So these good people, before leaving, wanting a more positive answer, again asked: "But, Padre, were you really present with us at that meeting?" And he replied: "Do you still doubt it?"

Saved from the Firing Squad

In a city of central Italy, a teacher and ex-secretary of a Fascist organization was accused of having furnished arms and bombs to the Fascisti, who perpetrated an explosion that killed military and civilians. But the teacher was innocent. Taken by force from her home to be tried and shot, she succeeded in bringing with her a rosary and a photograph of Padre Pio. They conducted her to observe the destruction and to see the spectacle of the dead whose death they attributed to her. They then brought her to the place of execution.

Meanwhile some members of the firing squad entered her home with the pretext of searching for arms. Instead they began stealing money, objects of gold, and clothing, until all of a sudden there thundered a shout, "Enough," so resolutely and imperiously that the soldiers fled, abandoning their loot.

The sister of the condemned girl, watching the whole scene cringing in a corner, recognized in the shouting of "Enough" the voice of Padre Pio.

The order to aim and fire had been suspended because of the arrival of an interminable column of armored cars, horses, cannon, ambulances and marching troops. The commander of the firing squad remained standing on a car as if hypnotized.

The young teacher could hardly breathe as she reflected that her hour would come when the last soldier passed by. She began to pray to Padre Pio for the grace of seeing God's Will in her execution. A gentleman approached her and asked what they had decided to do with her.

"I don't know, I no longer know anything, they are all away, there is only the commander there," as she pointed him out with an expression full of horror.

He was motionless, as if cemented to his post.

"Then consider yourself free and come with me." He brought her in his automobile to her home, where many women were comforting her grief-stricken sister. The condemned girl threw herself into her sister's arms, and then, taking a picture of Padre Pio from the wall, kissed it and pressed it to her heart. In that instant she felt a hand gently patting her cheek.

A few months later the teacher went to San Giovanni to thank her saviour.

"Padre," she said, "my life will not be enough to thank you." He said: "My child, how much your faith caused me to run."

Missions of Mercy

Mrs. Devoto of Genoa was seriously ill and in danger of losing her leg. A consultation was held, and the doctors decided to amputate the leg. One of her daughters was alone in her room praying that her mother would not have to submit to an operation. She also called upon Padre Pio for help. Suddenly she saw Padre Pio standing in her doorway, looking at her. Her desire to obtain the grace for her mother was so great that she did not stop to wonder how Padre Pio could be in Genoa, instead of San Giovanni Rotondo, hundreds of miles away; nor did she have the slightest doubt that he was actually there in person. Throwing herself on her knees she implored him, "Oh, Father, save my dear mother." He looked at her and said, "Wait for nine days." She wanted to ask him for an explanation; she raised her eyes, but saw only the door of her room—no light, no Padre Pio.

The next day she informed the doctors that they must wait. The doctors tried to convince her in vain. Even the other members of her family, when they saw that the mother was growing worse each day, could not dissuade her from her decision to wait nine days as Padre Pio had instructed her.

On the tenth day, when the doctors visited their patient, they were surprised to find the leg completely healed and their patient well on the road to recovery.

Mother, father, sons, daughters, in-laws, and grandchildren all came to thank Padre Pio for the grace which had been bestowed upon them, but Padre Pio would never accept thanks and would always say rather gruffly: "Go into the church and thank Our Lord and the Mother of Divine Graces."

Signora Concetta Bellarmini of S. Vito Lanciano declared that she was suddenly stricken with a blood infection followed by bronchial pneumonia with a very high fever. She was reduced to such a state that the doctors despaired of ever saving her. The flesh had become yellow from the infection which had spread throughout her body. A relative urged her to direct her prayers to Padre Pio. She prayed to him whom she had never seen, when suddenly in full daylight a stigmatized monk appeared to her, and smiling, blessed her without touching her as he stood in the middle of the room. The woman asked him if his appearance signified the grace of the conversion of her children, or else the grace of her physical cure. The Padre answered, "Sunday morning you will be cured"; then he vanished from the room, leaving an odor of perfume which the servant girl also smelled. After this visit her flesh turned a normal color, the fever ceased, and in a few days her health was completely restored. She went with her brother to San Giovanni Rotondo to see if Padre Pio was the one who appeared to her. When she arrived at the monastery and saw Padre Pio in the church, she turned to her brother and said, "There he is, he is the one."

Signor Arturo Bugarini of Ancona was urged by friends to turn to Padre Pio for the cure of his boy. Whilst he was standing near the bedside of his critically ill son, he felt three consecutive taps on his shoulder, whilst a voice said: "I am Padre Pio, I am Padre Pio, I am Padre Pio." At the same time he felt all over his body a wave of heat as if he were next to an intense flame, then all of a sudden it ceased. This visit of Padre Pio restored the health of the son. Father and son visited the monastery to thank Padre Pio for the miraculous cure and the spiritual conversion of the father and all the Bugarini family.

Emma Meneghello, a very pious young girl of 14, was afflicted with epilepsy which threw her into fits several times a week. One afternoon while in prayer, Padre Pio appeared to her and placed his hand on the bed sheet, then smiled and vanished. The cured epileptic arose to kiss the place where the Padre had placed his hand, and she noticed a cross of blood left on the sheet. A small square cut of the sheet with the bloodstains is conserved today in a glass picture frame. "Through the intercession of Padre Pio," she wrote, "I have obtained other favors, especially for dying babies."

Mrs. Ersilia Magurno, a woman of great faith, for two months was taking care of her husband, who was stricken with influenza. This illness would not have given alarm were it not that he was also afflicted with a very weak heart. Night and day, helped by a nun, the wife was watching her husband with every possible care, while praying and invoking Padre Pio. He grew worse, and the doctors advised that the last rites be given to him because of the very alarming failure of his heart. One night the wife noticed in the room a strong perfume of flowers. The next morning, however, a worse condition prevailed, and the dying man was approaching his end. A telegram begged Padre Pio's intercession. Two days later Mr. Magurno entered into a state of coma. The wife did not give up hope and sent a second telegram Finally on the 27th of February, 1947, the sick man, after a day of prolonged crisis, fell asleep. The nun was away, and the wife remained alone to watch him and at midnight, she noticed that his sleep was more restful than usual. At 7:30 in the morning, noticing that he was awakening, she rushed to his side and said, "How do you feel?" "I am cured, I am well. Padre Pio just left the room; open the window, please, and take my temperature." It was entirely normal.

"Ernesto," asked the astonished wife, anxious to hear and to know. "What are you saying? Have you seen Padre Pio? And what did he tell you?"

"He came together with another monk; he examined my heart and said, 'This fever will go away; tomorrow you will be cured, and within four days you can get up.' Padre Pio looked around, examined the medicines, read the medical reports and remained in the room all night." To confirm this miracle, a strong odor of violets was observed in the room.

Five months later, on July 27, the couple went to San Giovanni, and Mr. Magurno immediately recognized Padre Pio as the monk who had cured him. Padre Pio received him with fond greetings, and placing his hand on his shoulder said to him: "How much this heart has made you suffer."

We must not think that Padre Pio arrived always unexpectedly at the bedside of the sick, for some times he loved to announce the time of his coming. Once a little sick girl told her parents the approximate hour of his arrival, and the parents in their simplicity, not understanding the phenomenon, went to meet him at the railroad station. When they returned to the child disappointed, they found her asleep.

"Padre Pio didn't come," her mother said sadly, as soon as she saw her waking.

"Why, he was here up to now," the child answered. [10]

Another Pilot Without a Parachute

In my book "Fino alla Meta" I tell of a young aviator who was attached to a fighter squadron in the last war.

One day the lieutenant started off on a mission, and discovered right away that his plane was about to catch fire. He consulted his commanding officer by radio, who told him that if he could not put out the blaze he was to bail out of the plane with his parachute. All of his efforts were in vain so he jumped, but the parachute failed to open. He would have been killed had not a friar caught him in his arms and carried him to earth.

That evening he told his story to his commanding officer who did not believe a word of it, but gave him a short leave in order to recover from the shock of the experience.

When he reached home he told his tale to his mother. "Why it was Padre Pio, she said 'I prayed to him so hard for you!" and she showed him a picture of the Padre. Her son exclaimed: "Mother! That is the same man!"

The young soldier went to San Giovanni Rotondo to express his gratitude. Padre Pio said to him: "That was not the only time I saved you. At Monastir when your plane had been hit, I made it glide safely to earth." Which had actually happened. [11]

Padre Pio's Comments on His Bilocation

A few of Padre Pio's fellow-Capuchins and visitors dared to ask Padre Pio about his bilocation. Although reluctant to discuss the matter, he addressed the question on a few occasions.

How Does Bilocation Take Place?

One evening, Dr. Win. Sanguinetti (faithful friend and personal physician of Padre Pio) told us that he and a few others were in Padre Pio's room, when the doctor opened the following conversation:

Dr.: "Padre Pio, when God sends a saint, for instance like St. Anthony, to another place by bilocation, is that person aware of it?"

Padre Pio: "Yes. One moment he is here, and the next moment he is where God wants him."

Dr.: "But is he really in two places at once?"

Padre Pio: "Yes."

Dr.: "How is this possible?"

Padre Pio: "By a prolongation of his personality."

This explanation, obvious to Padre Pio, may be a problem which we leave to philosophers and theologians to explain. [12]

Do Bilocators Know They are Bilocating?

One day a monk who was speaking about the bilocation of St. Anthony of Padua who miraculously appeared in Lisbon, said to Padre Pio: "Perhaps these privileged of the Lord do not even know when their bilocation occurs."

Padre Pio quickly interrupting him as one who is experienced with such events answered:"Certainly they know. They cannot know if the body or the soul moves, but they are very conscious of what happens and they know where they are going."[13]

The Testimony of Padre Carmelo Durante

One day in the refectory we were talking of this and that.

I remember that in the conversation I was holding forth about a fact then unheard of: an aeroplane - I don't remember of which airline - had made the journey non-stop between Rome and New York in only six hours. To me and the others it seemed something incredible!

The Padre who until then had kept silent, interrupted in the middle and asked: "How long? How many hours, did you say?"

I answered, with increasing marvel: "Padre, six hours and what is more non-stop!"

The Padre also marveled over the fact but to the side exclaimed: "Six hours! Good heavens, but that is a long time! When I go it takes me only a second."

We asked him to explain himself, but he would say no more and only repeated: "I told you! I told you!"

"Six hours" in contrast to "a second" is indeed a long time.

But his mysterious "second" was that of a man in bilocation![14]

Being Present Where God Wants Us to Be

Padre Pio's fellow Capuchin and long-time assistant, Padre Pellegrino, teased him about his bilocation, telling him, "Padre, you cannot deny that every now and then you go to Paris, London, Berlin, etc., in bilocation, and at the same time you remain comfortably at San Giovanni Rotondo." Padre Pio responded: "If it is true that my person is in several places, by bilocation (…), you must ask God and not me. I can only tell you that I endeavor to remain always attached to the thread of His will. (…) I try to be always in one place: the one wanted by God."[15]

How do we know that all the accounts of Padre Pio's bilocation were accurate? By the nature of the case, bilocation belongs to the class of phenomena that we can only know about on the basis of witness testimony.

Such testimony is compelling in the present context because of the variety of the witnesses who have reported the Padre's appearances, the diversity of the circumstances in which these took place and the historical precedents of bilocation.

Yes, bilocation is an extraordinary claim but it is not the kind of claim that is likely to be invented by anyone wanting to concoct a plausible legend. And while one or two persons reporting the Padre's bilocation may be hallucinating, it is unlikely that the wide variety of people who reported his appearances via bilocation could each and every one be hallucinating. Many of them never even knew him or knew about him.

No one is compelled to believe in Padre Pio's powers of bilocation. But reviews of the bilocation reports do not indicate that the witnesses had any reason to "make up" their accounts. We should start off by giving them the benefit of the doubt. Then we should consider their reports within the framework of such other Padre Pio phenomena as the publicly visible stigmata as well as the credible accounts of bilocation in history. At that point, we can decide for ourselves if the claims are credible and therefore worthy of acceptance.

Padre Pio's bilocation is important not because of what it says about him. Rather, its importance lies in the insight it gives into the Power of God and the marvels of his creation.

Notes

[1]https://aleteia.org/2019/09/23/how-padre-pio-stopped-allied-forces-from-bombing-his-monastery-during-wwii/

[2]http://www.padrepio.catholicwebservices.com/ENGLISH/Bilo.htm

[3]Steven Pinker, *How the Mind Works* (New York: W.W. Norton, 1997), 564.

[4]Whitney Hopler, "Superhero Saints: Bilocation, the Power to Appear in Two Places," January 02, 2018 https://www.learnreligions.com/bilocation-appear-in-two-places-4000182

[5]https://www.gosanangelo.com/story/life/faith/2018/06/09/visit-agreda-revealed-more-lady-blue-mystery/684726002/

[6]Patricia Treece, *The Sanctified Body* (New York: Doubleday, 1989).

[7]"The Padre and Saint Therese," *The Voice of Padre Pio*, vol. xviii, no. 10, 1998.

[8]https://aleteia.org/2014/09/23/another-miracle-of-padre-pio-he-visited-cardinal-mindszenty-in-prison/

[9]"*Pray, Hope and Don't Worry*" newsletter, April 2008.

[10]Charles Mortimer Carty, *Padre Pio – The Stigmatist* (1963 by Radio Replies Press, St. Paul, Minnesota. 1973 by TAN Books), https://archive.org/stream/PadrePioTheStigmatistCartyCharlesMortimer6228_201810/Padre%20Pio%20the%20Stigmatist%20-%20Carty%2C%20Charles%20Mortimer_6228_djvu.txt

[11]Who is Padre Pio? (Rockford, Illinois: Tan Books, 1974), 9.

[12]Charles Mortimer Carty, op cit.

[13]http://www.miraclesofthesaints.com/2010/09/bilocation-of-st-padre-pio.html

[14]*The Voice of Padre Pio*, November 1998.

[15]Padre Pellegrino Funicelli, *Padre Pio's Jack of All Trades* (San Giovanni Rotondo, Italy: Our Lady of Grace Capuchin Friary, 1991), 26-7.

Relativity at Work – *"Seeing" the Future*

Another notable feature of the Padre Pio package was the Padre's all-but instinctive ability to "foretell" future events. Most of these cases involved dire happenings such as impending death. We have already seen this charism at work in some of his healing bilocations. As with bilocation, the charism was granted for serving a specific purpose in the divine plan. Most importantly, advance warnings enable spiritual preparation for our journey to eternity.

One of the oft-repeated "preparation for death" warnings involved a policeman who was guarding Padre Pio:

> Because of the great crowds that went to see Padre Pio, two policemen were assigned to the convent to protect him. One day in the sacristy after the celebration of Mass, while Padre Pio was disrobing the sacred vestments, he turned smiling to one of the policemen and said: "As soon I have given thanksgiving for the Mass and you are finished here, come to my room because I have to speak to you." The policeman was very happy for this invitation so, when he ended his service, he went to Padre Pio's cell. Padre Pio told him: "Listen to me! In not more than eight days you will die at your father's home, my son." The policeman answered him: "But Padre, I am feeling very well." But Padre Pio added: "Don't worry about it! You will be better in eight days. What is this life? A pilgrimage. We are on a train, my son! Ask your boss to go on leave to enjoy your family...you are going to die...and your relatives??? They do not know anything about this..." The

policeman, stunned by these words, asked: "Father, can I tell what you have told me?" "Not for now," the Father answered, "you can tell it only when you will be at home." The young man went to the town of St. Giovanni Rotondo and he asked for permission to go home. His boss did not want to grant him the permission because there was no suitable motivation. However, on Padre Pio's intercession, the policeman received the permission to go home. When the policeman reached his house, he told his parents: "I've come to greet you and Padre Pio has told me that I will die." After eight days the policeman died.[1]

Like bilocation, the "superpower" of looking into the future has been associated with other saints. St Anthony of Padua and St. Martin de Porres again were known for their prophetic gifts as was St. John Bosco.

As with bilocation, we wonder how such a capability is possible. How is it possible to know about an event that has not occurred yet? Modern science is again helpful in opening our minds to the possibilities before us. The Theory of Relativity, in particular, has shown us that our ordinary framework of time is not absolute.

What are relative in the Theory of Relativity are space and time. Space and time come into being with matter and motion. What we call space is simply a way of talking about relationships (e.g., location or distance) between different material bodies. Time is the measure of changes in the network of relationships we call space. Neither space nor time mean anything outside these relationships just as "stock market" does not mean anything outside certain buying-and-selling transactions (it is not a place).

According to Relativity, spacetime is the "relationship matrix" of causes and effects, agents and events.

Einstein's great insight dethroned Isaac Newton's ruling paradigm of independently existing absolute space and time. Yet, long before Newton and Einstein, the greatest thinkers of Christendom had gone far beyond both scientists in conceptualizing the relativity of space and time at an even more fundamental level. As physicist Stephen Hawking acknowledged in his *A Brief History of Time*, St. Augustine had famously said "that time was a property of the universe that God created, and that time did not exist before the beginning of the universe."[2]

But the discovery that time is a "property" of the Universe and not of God is simply the tip of the iceberg, a mere footnote to the far more radical theory of relativity embodied by the Judeo-Christian doctrine of God and *creatio ex nihilo*. This "theory" was carefully unpacked by Christendom's Triple A team of Augustine, Anselm and Aquinas and formally defined by the Fourth Lateran Council in 1215 A.D. In essence the Theological Theory of Relativity tells us that God is self-existence and that God's creation of the "world," the *creatio ex nihilo*, is a single eternally known act of both creation from nothing and constant conservation in being.

This means that time and history are simply part and parcel of the interacting web of causes and effects, agents and acts, events and processes, that constitutes the *creatio* and that is *known eternally* (not before, during or after) by God who is transcendent, timeless and infinitely perfect. Our entire space-time web, the histories of the Universe and humankind, are not concurrent with eternity, with God's transcendent Being. Rather, the *creatio* is one whole that is eternally known to God.

Hugh McCann writes, "God creates and is aware of all of history neither simultaneously nor at different times, but eternally."[3] Thus God knows all events and things "in a single timeless act of awareness that encompasses all of heaven and earth, in its complete history."[4]

How is this possible, we might ask? The response must be, why not? Take for instance the phenomenon of time dilation that is part of the theory of relativity. The clock used by a person who moves at different speeds will go slower than the clock of someone who is stationary. Then there is Einstein's Twin Paradox whereby a twin who travels at high speeds to a distant star ages at a slower rate than his brother on earth.

But these examples from physics pale beside the widely experienced phenomenon of precognition. This is the experience that thousands of people have had of "seeing" the future, so to speak. Many people have had the experience of knowing about some future event, for instance, a death and the manner in which it takes place, well in advance of its occurrence

– days and weeks, even years, before. It may be said that this kind of evidence is anecdotal and not scientifically systematic. But by its very nature it is not something that can be produced in a push-button manner. It is not a capability that is "in the genes" or susceptible to clinical trials. It is more like a gift freely given to certain people at certain times. A given person may have had three such experiences and never any other after.

Even more significant for Christians is the fact that prophecy is integral to the biblical revelation. Certain individuals, called prophets, were given visions of the future. These biblical instances are, of course, more authoritative and definitive than the accounts of precognition but in both sets of cases the underlying principle is the same: a future event has been *foreseen*.

The implications of these phenomena have rarely been considered. Let us grant that at least the accounts of biblical prophecies are reliable. And let us take only fulfilled biblical prophecies as our starting-point. How is it possible for someone to know about an event that has not taken place yet, about situations that do not yet exist? If we can really know the future, this means that the framework of our time-series can somehow by transcended.

The only possible conclusion is that there is a Source for whom there is no future, one who knows all of time and history eternally and reveals it to certain chosen individuals. Unbelievable it would seem – but then we have the hard evidence of knowledge of things that have not yet happened. No other explanation will suffice to explain this unthinkably mystifying state of affairs than the explanation that the Source is outside time. But this is precisely what we say about God, the Source of all things and the fullness of all perfection.

With all this in mind, we can better understand the ease with which Padre Pio peered "into" the future.

Viewed from the standpoint of God's eternity, we can see why the Padre Pio prophecies were possible. Some individuals are sometimes given the ability – by God – to "break out" of the infrastructure of everyday time.

We review here some of the reliably sourced and better-known accounts of Padre Pio's glimpses into the future:

Germany and the Second World War

Padre Pio's prophetic spirit as told by Padre Carmelo, who was Superior of the Convent of St. Giovanni Rotondo, is contained in this testimony: "During the last World War, we spoke of the war every day and of the uproarious military victories of Germany on all the fronts of battle. I remember one morning I was in the little parlor of the convent, and I read in a newspaper the news that the German troops were moving by now in the direction of Moscow. For me, it was important news; in fact, I saw in that news that the war would end with the final victory of Germany. Going out in the corridor, I met Padre Pio, and gladly I told him shouting: "Father, the war is ended! Germany has won!" - "Who has told you it?" asked Padre Pio. - "Father, the newspaper," I answered. - And Padre Pio said: "Germany has won the war? Keep in your mind that Germany, this time, will lose the war, worse than the last time! Remember it!." I told him: "Father, the Germans are already near Moscow." He added: "Remember what I have told you!" - I said: "But, if Germany loses the war, Italy will lose the war as well!" - And, he emphatically answered me: "We will see if together they will end the war." - I didn't understand those words, taking into account the alliance of Italy and Germany, but the words became clear the following year after the armistice with the English and Americans of September 8, 1943, and the Italian declaration of war against Germany.[5]

The Death of King George V

On January 20, 1936, Dr. Guglielmo Sanguinetti and several other men were visiting Padre Pio in his cell. As they were talking, Padre Pio suddenly interrupted the conversation and asked the men to kneel down with him and to pray. "We must pray for a soul who will soon appear before the judgment seat of God," Padre Pio said. When they were finished praying, Padre Pio asked his friends if they knew who they had been praying for. They replied that they did not know. Padre Pio told them that they had been praying for George V, the King of England. Dr. Sanguinetti told Padre Pio that he had just read in the paper that the King's health was not a cause for alarm. He had a head cold and no more. He was not in any danger. "What I am telling you is the truth," Padre Pio answered. About midnight, Father Aurelio heard a knock at his cell door. When he opened the door, there stood Padre Pio. "Let us pray for a soul who at this very moment has passed away and is now appearing before the tribunal of God. I am talking about the King of England," Padre Pio said. The two priests prayed together for a while. The next day, the newspapers announced that the King had died. His death occurred at the same time that Padre Pio and Father Aurelio were praying together for him. It was generally not Padre Pio's habit to make statements about political or world leaders. However, at the time of King George V's death, Padre Pio spoke of him and asked his friends to pray for his soul. [6]

The Election of Paul VI as Pope

After the death of Pope John XXIII, writes Bernard Ruffin,

> the friars in his monastery were pressing him to reveal who was going to be the next pope. Padre Eusebio was unusually persistent, and finally Padre Pio told him, "It's going to be Montini. Now will you be quiet?" Shortly afterward, the Archbishop of Milan, Cardinal Giovanni Battista Montini, was elected, taking the name Paul VI. Padre Pio's prediction, by itself, would not have been remarkable since many people were predicting Montini's election, except for the fact that, five years earlier, through Alberto Galletti – a former administrator of the hospital – Padre Pio had sent a message to Montini: "Tell the Archbishop that he will one day be Pope, and to be prepared. Mind you, tell him this." When Galletti relayed the message to Montini, the archbishop laughed and said, "Oh, these saints get some strange ideas!"[7]

Danger Ahead

Father Agostino of Campolieto was visiting Padre Pio in San Giovanni Rotondo on one occasion and mentioned to him that he would soon be going back to Algeria. Padre Pio told him that there was danger awaiting him there and advised him not to go. Father Agostino thought about Padre Pio's warning but did not feel that he could cancel his trip. When Father Agostino returned to Algeria there was a conflict that arose between the French people who lived in the country and some of the other ethnic groups. Because Father Agostino spoke French, he was considered a suspect. One night, at 11:30 p.m. the police came to his door and arrested him. At that moment, Father Agostino remembered Padre Pio's words. He was taken to the police station and questioned. He was finally released at 5:00 a.m. the next morning. The Capuchins in San Giovanni Rotondo noted that Padre Pio became suddenly ill at 11:00 p.m. the same night that Father Agostino was arrested. He remained ill until five o'clock the next morning.[8]

Change of Heart

In 1948 the Capuchin missionary Louis Magliacanni was about to return to India after a visit to his family in Italy. In few days the boat would leave Naples for India. A friend suggested him to visit Padre Pio. Louis said that Padre Pio was a fake. The friend suggested going anyway so that he had firsthand view of a fake. Padre Pio: "Padre Louis you will never go back to India. You will go to Arabia". Padre Louis: "Before coming here I thought you were crazy. Now I know you are." Padre Louis stayed for lunch at the friary. During lunch Louis was called on the phone. It was a cardinal from the Congregation for the Propagation of Faith. He told him not to go to India, and go instead immediately to Rome. In Rome the Cardinal and Padre Louis had a private audience with Pope Pius XII. The Pope told Louis: "You have been chosen to open a new mission in Arabia." Louis responded: "I knew that! Padre Pio told me. I thought he was crazy. Now I realize he is a saint!"[9]

His Own Death

Photographer Modesto Vinelli testifies that he used to see Padre Pio almost daily and take pictures of him. One day at the end of 1918 Padre Pio told him: "Modesto, we have fifty years ahead of us." He was also taking pictures on the day of the 25th anniversary, in 1943. Padre Pio told him: "Modesto remember that we still have 25 years to go." On September 20, 1968, Padre Pio told him: "Modesto, the fifty years are over." *Padre Pio died three days later. Modesto lived until 1983.*[10]

Notes

[1]https://www.padrepio.catholicwebservices.com/ENGLISH/knowledge.htm
[2]Stephen Hawking, *A Brief History of Time* (New York: Bantam, 1988), 8.
[3]Hugh J. McCann, "The God Beyond Time", *Philosophy of Religion – An Anthology*, edited by Louis P. Pojman (Belmont, CA: Wadsworth Publishing Company, 1998), 246.
[4]Ibid, 243-4.
[5]http://www.padrepio.catholicwebservices.com/ENGLISH/knowledge.htm
[6]https://padrepiodevotions.org/pdf/july-september2011.pdf
[7]C. Bernard Ruffin, *Padre Pio: The True Story* (Huntington, IN: Our Sunday Visitor, 2018), 430.
[8]https://padrepiodevotions.org/pdf/july-september2011.pdf
[9]https://caccioppoli.com/St.%20Padre%20Pio%20Prophecy.html
[10]https://caccioppoli.com/St.%20Padre%20Pio%20Prophecy.html

Multiverse – *A Social Network of Friends from the "Other World"*

Of Padre Pio it could be said that he literally had one foot in another world. This "other" world is the spiritual universe where the dead are alive again! Many of those who pass on to the after-life are not living "happily ever after." According to the ancient Christian vision, some may be in a state of endless agony and others may be paying off debts incurred while on earth before they are admitted into the fullness of joy. This latter is the state called Purgatory.

Padre Pio's teaching centered on the need to choose eternal union with God and to assist those detained in Purgatory, the "Poor Souls". So powerful were his prayers for the "poor souls" that many of them would come to thank him for his prayers prior to their final admittance into eternal glory. Thus his fellow monks could hear animated conversations in the Padre's cell. When mystified to find that no other physical person was present, they would be told that the conversation was with one or more of the Poor Souls who were stopping by on their way to their heavenly Home. They were literally the Grateful Dead.

Padre Pio's ministry to the Poor Souls was actually initiated by them:

> One day while praying alone, Padre Pio opened his eyes to see an old man standing there. He was surprised by the presence of another person in the room and explained in his testimony, "'I could not imagine how he could have entered the friary at this time of night since all the doors are locked."

Seeking to unravel the mystery, Pio asked the man, "Who are you? What do you want?"

The man responded, "Padre Pio, I am Pietro Di Mauro, son of Nicola, nicknamed Precoco. I died in this friary on the 18th of September, 1908, in cell number 4, when it was still a poorhouse. One night, while in bed, I fell asleep with a lighted cigar, which ignited the mattress and I died, suffocated and burned. I am still in purgatory. I need a holy Mass in order to be freed. God permitted that I come and ask you for help."

Pio comforted the poor soul by saying, "'Rest assured that tomorrow I will celebrate Mass for your liberation."

The man left and the next day Pio did some investigative work and discovered the veracity of the story and how a man of the same name died on that day in 1908. Everything was confirmed and Padre Pio celebrated a Mass for the repose of the old man's soul.

This was not the only appearance of a soul from purgatory asking Padre Pio for prayers. Pio claimed, "As many souls of the dead come up this road [to the monastery] as that of the souls of the living." Many times the souls would ask for a Mass to be said for them, highlighting the spiritual weight of a Mass and how it can lessen the time a person spends in purgatory before embracing the glories of heaven.[1]

Like most everything else, this dimension of Padre Pio testified to the matter-of-fact reality of the truths of faith in his life. What others tried hard to believe on faith, he took for granted. He not only "believed" these truths but he was interacting with the "facts" they spoke off on a daily basis. Padre Pio experienced many articles of faith first-hand! Of course, this was possible only because he had started off with humble and total faith, trust, obedience and commitment. He would have continued such a life even if he had not been given any supernatural experience. But he was granted extraordinary gifts precisely because of his total trust and consecration and his willingness to live with the constant suffering that came with the gifts.

With these gifts came his perception of the "other world" and its inhabitants. They came to him for his prayers and Masses because they too were aware of his perception of their world. When they "friended" him, he became a part of their social network! Despite being a suffering soul himself, he could never say "No" to those who asked his help in mitigating their suffering. Hence his constant concern for those undergoing the pains of Purgatory.

Life After Death?

Some might ask if there is life after death let alone Purgatory. Let us note that the overwhelming majority of human beings across the course of history have taken it for granted that death is not the end, that the human person in some fashion lives to tell the tale, that there is a life after death. This striking and seemingly instinctive belief has been embodied in the religious traditions and philosophical reflections of most cultures. In some instances, it has claimed confirmation in a series of experiences and encounters.

The picture above is the one, in different hues and colors, that was painted by the primordial peoples of the earth as also ancient India, China and Persia and, finally, Judaism and Christianity. Even the most ancient peoples believed in a Heaven, a Hell and a domain in-between. The Vedic Hindus did not know anything of reincarnation and taught that after death you enter a permanent state of joy or sorrow (The World of God or the House of Clay). The ancient Persians, Egyptians, Mesopotamians, Greeks, the primordial tribes, the Chinese, the Japanese, the Mesoamericans, the Jews all believed in an after-life where one's deeds in this life affected one's destiny in the next. In classical Christianity we find a highly structured description of the after-life. We have the Holy Ones now united with God (Heaven), those who are being purified in preparation for divine union (Purgatory) and the damned who are separated from God (Hell). There is also a resurrection of the body at the end of history.

One radical development in the Christian dispensation was the proclamation that the sacrificial death of Jesus made it possible for those who chose God to go to Heaven. This idea was radical in one sense and yet strangely continuous with the intuitions in some religious traditions of a sacrificial Savior (Hinduism, Judaism and Zoroastrianism in particular). Although some Christians have said that only professed Christians can go to Heaven, Christian writers as far back as St. Paul and St. Augustine had said that Heaven is open to all who trust in God and do his Will (see also the teaching of Jesus in *Matthew* 25). St. Paul said in fact that God wills the salvation of all. At the same time Christian teaching was clear that the reality of human freedom made it possible to reject God. It held that those who do so will be eternally separated from him.

Another element of the Christian vision is the claim that Jesus of Nazareth was not only crucified but also rose from the dead. The proclamation of his resurrection is central to Christianity and, for Christians, serves as an enduring foundation for the reality of life after death.

By far the most compelling feature of the claim of the resurrection of Jesus is the transformation of the followers of Jesus and the genesis of the Christian movement. What transformed eleven fearful peasants and fisher-folk into superheroes who preached the Good News across the world despite trials and tribulations and eventually horrendous deaths? What galvanized them to take on the most powerful empire of the day? The hallucination hypothesis is not remotely plausible for those who know the causes and characteristics of hallucinations (an individual experience, drugs, mental illness, an expectation of seeing that which is allegedly witnessed – none of which apply in the present case). The idea of a hoax is wildly implausible given the improbability that anyone would embrace a gruesome end in order to perpetrate a fiction. And the notion that a transplanted myth was the impetus behind the conversion and martyrdom of a ragtag band of one-time cowards and whiners stands self-condemned simply in the telling. It seems undeniable that the extraordinary transformation could only be explained by an extraordinary event and, in this respect, the Resurrection makes perfect sense.

Ludwig Wittgenstein, the most influential philosopher of the last 100 years, speaking as a modern Jewish witness to the Resurrection, said, "What inclines even me to believe in Christ's resurrection? ... If he did not rise from the dead ... We are in a sort of hell ... cut off from heaven. But if I am to be REALLY saved, – what I need is CERTAINTY – not wisdom, dreams or speculations – and this certainty is faith. And faith is what is needed by my HEART, my SOUL, not my speculative intelligence."[2]

Now this picture of a continuum between the here-and-now and the hereafter was torn to shreds by the modern era. Then came the 1970's and the startling reports of clinically dead subjects who reported life on the other side of death. By the 1980's, according to a Gallup poll, over 8 million in the US alone reported NDEs (one such subject happened to be a world-famous atheist from the UK); in the early 2000s, according to researcher Nancy Bush, the number of NDE subjects in the US had been estimated to reach 22 million. There were positive NDEs and negative NDEs (also called "distress" NDEs) – shades of Heaven and Hell! Erlendur Haraldsson, a veteran researcher in this area, noted that "Recent survey data indicate a widespread belief in life after death in most countries of Western Europe and North America. ... Some persons not only believe in life after death, but they also report encounters with the dead."[3]

The *Annals of the New York Academy of Sciences* reported in February 2022 that "Scientific advances in the 20th and 21st centuries have led to a major evolution in the understanding of death. (...) A multidisciplinary team of national and international leaders (...) examined the accumulated scientific evidence to date." Their report

> "represents the first-ever, peer-reviewed consensus statement for the scientific study of recalled experiences surrounding death. Among their conclusions:
>
> 1. Due to advances in resuscitation and critical care medicine, many people have survived encounters with death or being near-death. These people -- who are estimated to comprise hundreds of millions of people around the world based on previous population studies -- have consistently described recalled experiences surrounding death, which involve a unique set of mental recollections with universal themes.
> 2. The recalled experiences surrounding death are not consistent with hallucinations, illusions or psychedelic drug induced experiences, according to several previously published studies. Instead, they follow a specific narrative arc involving a

perception of: (a) separation from the body with a heightened, vast sense of consciousness and recognition of death; (b) travel to a destination; (c) a meaningful and purposeful review of life, involving a critical analysis of all actions, intentions and thoughts towards others."[4]

Purgatory?

In the ancient world, and in Christianity through the centuries, there was a constant awareness of the "connection" between the living and the dead. Central to the fabric of the human reality as they understood it was:

- the intercession of the saints for those on earth,
- the travails of the souls in Purgatory,
- the appearances of these saints and "Poor Souls" to those on earth,
- journeys made by the living to the "other world."

The world of "departed" souls was just as real as the physical world. Heaven and earth, the blessed dead and the living – it was all an interconnected web, a web far more enduring and extensive than the web of life. We are surrounded by a "great cloud of witnesses," said the Epistle to the Hebrews.

Most Christians accept Heaven and Hell as the two ultimate options for every human person. But, ever since the Reformation, some have rejected the whole idea of an intermediate state for those who have been "saved" and still need to be "purified." This is the state often called "purgatory" ("purge" meaning "purify"); it is further affirmed that the souls in Purgatory can be assisted, by the prayers of the living.

In considering these notions in more detail we should recognize the following:

- Purgatory and the intercession of the saints were taken for granted as true by Christians from the very beginning in their faith and applied in their devotional practices and liturgies.
- Both were taught by the Fathers of the Church on the basis of their interpretation of Scripture.

- Eminent Protestant Christians like C.S. Lewis have accepted Purgatory and prayers for the dead.

What the First Christians Believed

The belief that many of the dead undergo a purifying process before entering Heaven was instinctively adopted by the first Christians. Accordingly they prayed for their deceased friends. Catherine Marshall, a well-known Protestant writer, has chronicled this historical detail that is unknown to many today:

> "Did Christians then [in the first centuries of Christianity] know something about personal contact across the dividing line of death that we do not know today?
>
> "The first fact I uncovered was that, from the establishment of the church after Pentecost to the Protestant Reformation in the sixteenth century, *prayer for and with the dead had been universally practiced by Christendom.* This followed naturally from two bedrock beliefs of the Apostolic Church: (1) that of immortality, that Christ – through His resurrection, to which the earliest Christians claimed to be eye-witnesses had for ever conquered death; and (2) that the unity of the church – the Body of Christ – was such that death could not dismember it.
>
> "The prayers of the first Christians for the dead were not mere petitions that the departed from some state of eternal punishment. To them, prayer was as natural as breathing. It was the life of the spirit; it was their lifeline with the Risen Lord. He had taught them that hey were 'members one of another' of an organic fellowship. ... To them it was unthinkable that the incident of death should sever their communion with each other and with the living Christ. ...
>
> "The writings of the earliest church fathers who came after the apostles – such as Tertullian and Cyprian – also bear unanimous testimony that the fellowship of prayer with those in the next life was taken for granted. Then in Augustine's Confessions we have his prayer for his mother Monica, written after her death."[5]

Freedom and Salvation

The question is why did the Christian community believe in this and how did the belief fit in with their Scriptures and doctrines? On the question of faith we should point that free choices are critical to salvation. The standoff between those who proclaim that faith alone is sufficient for salvation and those who allegedly point to faith and works or works alone as the path to salvation rests on a false dilemma. The New Testament clearly teaches that freewill is involved in salvation.

The Gospels teach that one's free decisions determine one's eternal destiny: "For the Son of Man is going to come in the glory of his Father with his angels, and, when he does, he will reward each one according to his behavior." (*Matthew* 16:27). This is a teaching that is reiterated in the rest of the New Testament. Says St. Paul, "Your stubborn refusal to repent is only adding to the anger God will have toward you on that day of anger when his just judgments will be made known. He will repay each one as his works deserve. For those who sought renown and honor and immortality by always doing good there will be eternal life; for the unsubmissive who refused to take truth for their guide and took depravity instead, there will be anger and fury." (*Romans* 2:5-8).

But is the choice we make for God a kind of work and does this mean that salvation comes from works? The reality is that God is omnipotent and sovereign. The only real power that humans have, a power not shared by animals, is a negative power – the power to say No to God. The human person can reject the divine offer of salvation. But the person who chooses NOT to say No is not performing a work. In not saying No, you are simply choosing not to exercise the negative power at your disposal.

So what of faith? In the Christian account, faith means surrendering your will to Jesus, trusting him and following him however imperfectly. In the Christian view this is the ordinary way to salvation but (as noted) God is not limited by his ordinary way in bringing his creatures to salvation.

Purgatory and the New Testament

As for Purgatory, the New Testament says clearly "nothing defiled" shall enter Heaven. (Revelation 21:27). Most people who die have some defilement or other when they leave this world. Does this mean they go to Hell? Some Christians might say that their faith in Christ will save them no matter what their imperfections. But here it is not a question of salvation but of whether you can enter Heaven while defiled. The plentitude of perfection that is God cannot, by the very nature of the divine Nature, be united with anything that is in any way impure. Even those who have faith in Christ cannot claim to be fully transformed as persons. Even if you had made the choice for Heaven, your soul still has to be re-molded to live in Heaven.

The New Testament hints in a few passages that there is a possibility of some kind of punishment after death that is not final. St. Paul writes, "For no one can lay a foundation other than the one that there is, namely, Jesus Christ. If anyone builds on this foundation with gold, silver, precious stones, wood, hay or straw, the work of each one will come to light, for the Day will disclose it. It will be revealed with fire, and the fire will test the quality of each one's work. If the work stands that someone built upon the foundation, that person will receive a wage. But if someone's work is burned up, that one will suffer loss; the person will be saved, but only through fire." (1 *Corinthians* 3:11-15). The person is saved but only through fire.

In his book *Eternal Security*, the well-known Baptist writer Charles Stanley has an interpretation of this verse that is of interest here. He asks us to visualize the believer standing before God. "Picture yourself watching saint after saint rewarded for faithfulness and service to the King—and all the time knowing that you had just as many opportunities but did nothing about them. We cannot conceive of the agony and frustration we would feel if we were to undergo such an ordeal; the realization that our unfaithfulness had cost us eternally would be devastating. And so it will be for many believers. Just as those who are found faithful will rejoice, so those who suffer loss will weep. As some are celebrated for their faithfulness, others will gnash their teeth in frustration over their own shortsightedness and greed. We do not know how long this time of rejoicing and sorrow will last. Those whose works are burned will not weep and gnash their teeth for eternity. At some point we know God will comfort those who have suffered loss (see Rev. 21:4) …On the other side of the coin, we can rest assured that none of our good deeds will go unnoticed, either."[6]

As a Baptist, Dr. Stanley does not, of course, believe in Purgatory. But clearly his interpretation of this verse in Corinthians entails the idea of some kind of "loss" after death for those who still end up in Heaven.

Other verses have been used in support of the doctrine. These include Jesus' reference to a sin that "will not be forgiven either in this world or in the next" (*Matthew* 12:32) (which implies the possibility of some kind of restitution in the next world) and *Matthew* 5:25: "Amen, I say to you, you will not be released until you have paid the last penny."

Fathers, Councils and Liturgies

All these verses make sense in the light of Purgatory. It should be remembered that such fundamental doctrines of Christianity as the Trinity were developed not by assembling proof-texts but by interpreting the teaching of Christ through the lens of the Fathers, Councils and liturgies. This is the true birth-place of Purgatory and other basic beliefs of the early Christians.

Purgatory as Part of Sanctification

Much of the revulsion for Purgatory is a result of misconceptions about the teaching. As the Anglican theologian John Macquarrie has said, Purgatory is "one aspect of that process of sanctification, whereby we are conformed to Christ." It is a completion of earthly penance and of conversion. The pains of Purgatory, as Francis Mannion puts it, are pains not of punishment but of growth and transformation, of putting on a new self. In Macquarrie's words, it is "the painful surrender of the ego-centered self that the God-centered self of love may take its place." The fire of Purgatory is a cleansing fire.

Detox Center

Purgatory is also a recognition that God always respects our freedom. Let's say we freely choose eternal union with God. Next, we have to address the self-created impediments that come in the way of consummating our union with him. The state of the soul cannot be transformed by a magic wand: it has to be cleansed with our full consent. The soul has to be "straightened out," rehabilitated, detoxified. And this is what is meant by Purgatory. It is a detox facility, a rehab center. As with any such facility, depending on the degree of the patient's problem, the kind and duration of the therapy will vary. The patients cannot help themselves. But they can certainly receive assistance from outside.

Dr. Johnson on Purgatory

The famous Dr. Samuel Johnson helped clarify some of the misconceptions about Purgatory to his biographer Boswell: [Christians who affirm Purgatory] "are of the opinion that the generality of mankind are neither so obstinately wicked as to deserve everlasting punishment, nor so good as to merit being admitted into the society of the blessed spirits; and therefore that God is graciously pleased to allow of a middle state, where they may be purified by certain degrees of suffering. You see, Sir, there is nothing unreasonable in this."

Perhaps the most memorable Protestant defense of Purgatory came from C.S. Lewis:

> "Of course I pray for the dead. The action is so spontaneous, so all but inevitable, that only the most compulsive theological case against it would deter me. ...
>
> "I believe in purgatory...
>
> "Our souls *demand* purgatory, don't they? Would it not break the heart if God said to us, 'It is true, my son, that your breath smells and your rags drip with mud and slime, but we are charitable here and no one will upbraid you with these things, nor draw away from you. Enter into the joy'? Should we not reply, 'With submission, sir, and if there is no objection, I'd *rather* be cleaned first.' 'It may hurt, you know'—'Even so, sir.'"
>
> "I assume that the process of purification will normally involve suffering. Partly from tradition; partly because most real good that has been done me in this life has involved it. But I don't think the

suffering is the purpose of the purgation. I can well believe that people neither much worse nor much better than I will suffer less than I or more. . . . The treatment given will be the one required, whether it hurts little or much."[7]

The Pains of Purgatory

In what consists the pain of Purgatory as traditionally understood? Frank Sheed writes, "We cannot pretend to know what the suffering of a disembodied soul is, but for the souls we are considering there can be no question what the principal element is. They long to see the unveiled face of God, yet they could not bear to see it while any uncleanness remains in them. They suffer from the anguish of their desire and the clear vision of the taint of self still in their own will. By accepting God's will, they find healing for their own will. ... When the cleansing is complete, they are at last fully human. The evil they have done is purified and they can face God and both those who have wronged them and those they have wronged."[8]

Padre Pio and Purgatory

Padre Pio did not simply believe Purgatory. It was an everyday part of the multiverse in which he lived as illustrated by a multitude of incidents:

The Soldiers

One night in 1944 the friars heard loud voices coming from downstairs saying **"Viva Padre Pio"**. The superior Padre Raffaele da S. Elia a Pianisi told the doorkeeper Fra Gerardo da Deliceto to let those people out and lock the door properly. Fra Gerardo went downstairs, didn't find anybody, and the door was double locked as it was supposed to.

He went back to report. Padre Raffaele was puzzled and went straight to Padre Pio asking if he knew something.

"Oh! Those were soldiers who had died on the battleground, and came to thank me."

Conversation with Unseen Visitors

A friar testified: We were all in the dining room when Padre Pio got suddenly up and walked at steady pace to the door of the convent. He opened it and started having a conversation.

The two friars that went with him didn't see anybody and started thinking that something might be wrong with Padre Pio. On the way back to the dining area Padre Pio explained:

"Don't worry. I was talking to some souls on their way from Purgatory to Paradise. They came to thank me that I remembered them today in the Mass."

A Mystery Known in Paradise

Maria Pompilio, when her brother died, asked Padre Pio to intercede so that he could come in her dreams. Her brother came in a dream and said:

"Padre Pio assisted me in my agony. He stayed until the Judge judged me. I was given eleven years of Purgatory, but [because of the] intercession of Padre Pio the pain was reduced to one year.

The morning after Maria went to see Padre Pio. From afar he said: "Are you glad now?" "Yes, it seems that you are everywhere."

"What sense has for me being on earth if I can't go up and down. I know.

Your brother told you that the mystery of my life will be known only in Paradise."[9]

About Padre Pio and the Poor Souls, his personal assistant Fr. Alessio Parente writes, Purgatory was "something Padre Pio knew about! When he spoke of the Poor Souls and of what they suffer, he spoke from personal experience, and consequently he did all in his power to instill in all those who came into contact with him love for those poor unfortunate brothers and sisters of ours. He made us understand that it is not enough for us to be moved with pity toward them, but we must take action. We must follow his example by offering our daily prayers and sufferings to the Lord for their release."[10]

Notes

[1]https://aleteia.org/2017/11/07/when-padre-pio-was-visited-by-a-soul-from-purgatory/

[2]Ludwig Wittgenstein, *Culture and Value*, transl. Peter Winch (Oxford: Blackwell, 1980),33.

[3]Erlendur Haraldsson, "Survey of Claimed Encounters with the Dead," *Omega*, Vol. 19(Z), 1988-89.

[4]https://www.eurekalert.org/news-releases/948999

[5]Catherine Marshall, *To Live Again* (London: Collins, 1977), 218-9, 221.

[6]Charles Stanley, *Eternal Security – Can You Be Sure?* (Nashville: Oliver Nelson, 1990),
127.

[7]*Letters to Malcolm, Chiefly on Prayer* (London: Collins, 1966), 109-111.

[8]F.J. Sheed, *Where Will You Spend Eternity?* (London: Sheed and Ward, 1977), 57-8.

[9]https://caccioppoli.com/Close%20encounters%20of%20Padre%20Pio%20with%20deceased%20souls%20in%20Purgatory,%20Guardian%20Angel,%20devil.%20His%20own%20words%20on%20temptation,%20virtues,%20life..html

[10]Alessio Parente, *The Holy Souls* (San Giovanni Rotondo, Italy: Padre Pio of Pietrelcina Editions, 1994), 63.

Blood-biome –
The Wounds of Destiny

"I bear the marks of Jesus on my body." Galatians 6:17

Padre Pio bore on his body the wounds of the crucified Christ, what are called the stigmata. These were the wounds inflicted on Jesus of Nazareth by the piercing of his hands and legs and his side. "Stigmata" has the same root as "stigma": it was a "mark of disgrace." Like crucifixion, it comes from a practice pioneered by ancient Rome: runaway slaves were branded like cattle and the mark left on them was called "stigma." For authentic stigmatists like Padre Pio, their stigmata were painful marks of shame.

Initially, starting in 1915, the wounds were felt but not materially visible. But they became physically present in 1918 and were visible till shortly before his death in 1968. Blood gushed freely from his hands and as a result they were always bandaged. Since Church law at the time required bandages to be removed during the liturgy of the Eucharist, the wounds were publicly visible when Padre Pio celebrated Mass.

"On Friday, the 20th of September 1918, there happened to him an event that not only changed his whole life, but that singled him out from the rest of humanity. He was praying in his stall in the choir when suddenly the monks heard a piercing cry. On running to find the cause of it they came upon Padre Pio lying unconscious on the floor of the choir, his hands, his feet and his side marked with deep, bleeding wounds. He was carried to his cell where he gradually recovered consciousness, begging his brothers to keep his secret. He had worn invisible stigmata for three years, and now they were there for all to see. . . . He has been the subject of endless and often painful medical examinations, and has undergone every kind of supposedly healing treatment, but the wounds remain open and completely free from infection. He loses about a cupful of blood every day from his side, which is covered at all times with a linen cloth to prevent the endless staining of his garments. He wears brown half gloves on his hands excepting when he is saying mass."[1]

The stigmata were excruciatingly painful and Padre Pio's slow gait was caused by the open bleeding wounds on his feet. He lived with constant pain. He did not ask for the stigmata. He was, in fact, ashamed of them and embarrassed about showing them even to his fellow monks. But he bore them in obedience to his divine Master no matter what the cost. They were given by Jesus to Padre Pio in order to assist in the mission of salvation like others before him as St. Paul explains in *Colossians*.

When the time came for the end of Padre Pio's earthly mission the stigmata disappeared since that phase of his mission was now complete. In any case, there is no suffering involved if blood flows from a corpse!

The stigmata played another role in the mission of this living martyr. They served as a "sign" of the supernatural – a sign that remained stubbornly present in the face of every evasive maneuver. They could not be explained away as anecdotal. They were literally in your face. The skeptics' maneuvers were many but, as is historically evident, all failed.

The Reason for the Stigmata

So how do we explain the stigmata?

Scientific theories are basically frameworks which try to explain why certain things are the case. The scientist assembles the available facts and then tries to explain them in terms of the theory. When it comes to the stigmata, we start with the fact that the appearance of these "signs" is not an isolated occurrence.

Central to the life of Jesus was his Passion and death on our behalf. As if to underline its importance, the climactic moments of the Passion were literally incarnated in the lives of over 300 persons across Christian history with over 60 of them canonized or beatified. That is to say, the wounds of Jesus Christ, called the Sacred Stigmata, mysteriously appeared on their bodies – on their hands, legs, sides or heads - in all their painful reality.

In more recent cases, the wounds were verified by medical authorities although their origin was inexplicable on a natural level. The wounds caused agonizing pain and those who bore them were often publicly humiliated. The stigmatists themselves told us why they were chosen. They were asked to offer up their suffering to the Lord in amendment for their sins and the sins of all who had turned from God.

When we unite our sufferings with those of Christ, God releases great spiritual power through them.

With his death and resurrection, Jesus redeemed the human race "once for all." But the fruits of this redemptive act have to be accepted by every human person to be effective in their lives. And those who are followers of Jesus are called to assist in the application of these fruits as memorably described by St. Paul. All of the authentic stigmatists willingly, even enthusiastically, accepted their part in bringing the Redeemer's work to the world at large – much like the Apostles themselves who gave their lives to the spreading of the Gospel.

St. Paul was perhaps the first stigmatist ("I bear on my body the marks of Jesus."). Below is a brief overview of the best-known stigmatists taken from the book *They Bore the Wounds of Christ: The Mystery of the Sacred Stigmata* by Michael Freze.

St. Gertrude the Great, 1256-1302	Infant Jesus; Jesus in glory; stigmata
St. Catherine dei Ricci, 1522-1590	Suffering of Jesus; stigmata
St. Rose of Lima, 1586-1617	Jesus as a child; stigmata
St. Francis of Assisi, 1182-1226	Received the wounds of Christ on September 14, 1224. The wounds remained on him until his death.

St. Clare of Montefalco. d.1308		In addition to bearing the stigmata, an imprint of the wounds was found on her heart after her death.
St. Catherine of Siena, 1347-80		Received the stigmata in 1375. At her request to the Lord, the wounds later disappeared but re-appeared at her death.
St. Lidwina of Holland	1380-1433	Received the stigmata in 1407 and carried it for the rest of her life.
St. Rita of Cascia	1381-1457	A wound from the crown of thorns on forehead. Received 1441.
Bl. Osanna Of Mantua	1449-1505	After she requested the Lord to share in His Passion, she received first the wounds of the crown of thorns and then the five wounds.
St. Teresa of Avila	1515-82	Recipient of transverberation, a stigma on the heart.
St. Rose of Lima	1586-1617	Received the Five Wounds.
St. Margaret Mary Alacoque	1647-90	Received the stigmata invisibly.
St. Veronica of Giuliani	1660-1727	Received the wound of Christ's heart in 1697.
St. Mary Frances of the Five Wounds	1734-91	Received the stigmata on hands, feet and side.
Anne Catherine Emmerich	1774-1824	Stigmata appeared in 1812 after she asked for a share in the Passion. Her wounds were medically examined but no natural cause for them could be found.

Louise Lateau	1850-83	The Five Wounds and later the wounds from the Crown of Thorns. Her wounds were examined by medical authorities.
St. Gemma Galgani	1878-1903	Received the stigmata which remained invisible at her request.
Sr. Josefa Menendez	1890-1923	Received the stigmata and underwent severe sufferings.
Sr. Faustina Kowalska	1905-38	Invisible stigmata
Berthe Petit	1870-1943	Invisible stigmata
Alexandrina Da Costa	1904-55	Received the stigmata
Theresa Neumann	1898-1962	Bore the stigmata for 36 years. She carried all the wounds of the Passion: hand and foot wounds, wound to the heart, shoulder wound, 30 scourge marks and 9 head wounds.
Marthe Thorns Robin	1902-81	In 1930 she received the Five Wounds and the Crown of wounds.

Although he was not the first or the last stigmatist in history, Padre Pio was the first priest to have manifested the stigmata (St. Francis of Assisi was not a priest). His priestly vocation was particularly relevant because of the amplified agony he underwent while celebrating Mass. This is significant because the eucharistic celebration, it should be remembered, is a sacramental and bloodless participation in the sacrifice of Calvary – a participation and not a repetition. St. Paul pointedly notes: "The cup of blessing that we bless, is it not a participation in the blood of Christ? The bread that we break, is it not a participation in the body of Christ?" (1 *Corinthians* 10:16). The Mass is equally a participation in the Son's eternal intercession before the Father: "he lives forever to make intercession for them." (*Hebrews* 7:24-25). This bloodless sacrifice is performed at the express command of Jesus: "This is my body, which will be given for you; do *this* in memory of me." (Luke 2:19)

What we have before us are the following: the crucifixion of Jesus, the scriptural call to participate in the suffering of Jesus and the fact of individuals in different places and times who appear to bear the wounds of the crucified Jesus. This is the framework or "theory" within which we should consider the claim of the stigmata in the life of Padre Pio.

In recent years, the Vatican released the records of the first major ecclesiastical investigation into the origin of the stigmata. This is what the newly released record tells us:

> In an article published by the L'Osservatore Romano, Francesco Castelli, a biographer of St. Pio of Pietrelcina, has revealed details of the first investigation in 1921 by the Holy Office—now the Congregation for the Doctrine of the Faith—into the life of the Italian priest and the authenticity of his stigmata. (...)
>
> In 1921, the Holy Office charged Bishop Carlo Raffaello Rossi, who would later be made cardinal, with visiting Padre Pio to investigate his life and the origin of his stigmata. In his report, Bishop Rossi wrote that Padre Pio "held his head high and was serene, his look lively and sweet, his words gleamed with kindness and sincerity."
>
> The task that began on June 14 of that year lasted for eight days, during which Bishop Rossi observed Padre Pio in detail. He wrote that he was very gentle with his brothers, beloved by his superiors for being a "great example and not a gossiper." He spent 10-12 hours a day in the confessional and he celebrated the Mass with "extraordinary devotion."
>
> Bishop Rossi said the observations were not sufficient and he decided to interview Padre Pio, who responded to 142 questions under oath with his hand on the Gospels. Castelli says his answers almost constituted a complete biography.

Questions such as, "Who gave you the stigmata? For what reason? Were you given a specific mission?" were answered serenely by the Italian saint in the following manner:

"On September 20, 1918 after celebrating Mass, while I was giving thanks in the Choir, I was repeatedly overcome by trembling. Later I became calm again and I saw our Lord as if He were on the cross—but I did not see if He did have a cross—lamenting the lack of response from mankind, especially from those consecrated to Him who are His favorites. He was showing that He was suffering and that He desired to unite souls to His Passion. He invited me to enter into His sufferings and to mediate upon them: and at the same time to concern myself with the health of the brothers. Immediately I felt full of compassion for the sufferings of the Lord and I asked Him what I could do. I heard this voice: 'I unite you to my Passion'. And immediately, the vision having disappeared, I came to and I saw these signs from which blood was flowing. I did not have them before."

Castelli said Bishop Rossi went even further. He asked to examine the wounds and as he did so he asked Padre Pio about them. He saw that the wound in his side "changed aspects frequently and at that moment was in the shape of a triangle, never before seen. Regarding the wounds of Padre Pio, he gave me precise and detailed answers explaining in addition that the wounds in his feet and side had a sort of radiant aspect."

Bishop Rossi concluded that the wounds were not "the work of the devil" nor were they the result of "deceit, fraud or a malicious or evil ability. Much less were they the result of external suggestion, nor do I consider them to be the result of suggestion." The distinctive elements "of true stigmata were found in those of Padre Pio," he added. Other details such as his high fevers and perfume-like scents confirmed the veracity of the phenomenon.[2]

Investigations and "Explanations"

We have already seen why some of the routes taken by stigmata skeptics reach a dead-end when applied to the Padre Pio phenomenon.

By far the most comical "explanation" is the charge that it was generated by the application of carbolic acid. This charge got a recent re-run because the Luzzatto book references a pharmacist who had given Padre Pio carbolic acid and veratrine in 1919. Luzzatto himself expresses no opinion on the origin of the stigmata. But some of the sensationalist media outlets seized on the reference to run headlines such as this from the *Daily Mail*, "Italy's favourite saint Padre Pio 'used carbolic acid to self-inflict stigmata', claims historian."

The charge is asinine because, although it was often used as a disinfectant in the early twentieth century, carbolic acid is known to cause burns and would be toxic at the levels at which Padre Pio is supposed to have used it. Even as far back as 1931, the British Medical Journal observed that "the carbolic acid group constitutes one of the most popular suicidal poisons."[3] A modern medical site notes:

> When administered in diluted strength, it is particularly useful as a topical anesthetic, and can relieve toothache pain. When applied in more concentrated formulations, however, carbolic acid can behave as a poison, in which case it targets the nervous system of the human body with dire effects.[4]

In a thorough and masterful response to the chemical agent claims, Frank Rega writes:

> The charge that St. Padre Pio used chemicals to sustain the stigmata was shown to be baseless as far back as 1919, less than a year after their occurrence. In that year, a little-known attempt at a medical "cure" effectively ruled out chemical agents as the cause of the wounds on the saint's hands, feet and chest.
>
> Only three official medical examinations of St. Pio's wounds were ever authorized, and they all occurred in 1919. The examining physicians did their work independently of the others. One of the

three was Dr. Amico Bignami, Professor of Pathology at the Royal University of Rome (Regia Università di Roma). The Procurator General of the Capuchin Order, responding to a request from the Holy Office at the Vatican, invited Dr. Bignami to the friary at San Giovanni Rotondo. He was asked to perform a medical examination on the wounds of Padre Pio, and to render his judgment.

Bignami was an atheist and logical positivist, which meant that he would only give credence to phenomena which could be proven scientifically or explained naturally. At first he refused the commission, but upon learning that the request originated from the Curia, he accepted, but wished no one to know of the assignment.

The Examination

He arrived in July of 1919, accompanied by the Capuchin Provincial, Padre Pietro of Ischitella. He only remained a few days, and drew up his report on the 26th of the month. During his short stay he examined Padre Pio several times.

To his credit, in his final report Bignami depicts the friar as having an expression on his face that is "full of goodness and sincerity, which inspire affection ... that in spite of his apparent weakness he supports fatigue very well. He can, for example, hear confessions for even 15-16 continuous hours without eating. He usually eats little . . . He has a vivacious, gentle, and sometimes wandering look..." (9)

His five-page report then proceeds to describe in detail the anatomical and histological characteristics of Padre Pio's wounds (10). He is struck by the symmetry of the "lesions," that is, that the wounds on the palms and soles are in a corresponding place to those on the opposite sides of the same hands and feet. As for the chest wound, he considers that it is simply the result of a superficial abrasion of the epidermis.

He notices that the skin around the wounds is colored with a halo from tincture of iodine, and this arouses his suspicion. In response to Bignami's questions about the iodine, Padre Pio says he uses it as a disinfectant a couple of times a week or more, and it also helps to lessen the bleeding. (...)

As a medical professional Bignami believed that Padre Pio's lesions should respond to clinical treatment. In order to conclusively demonstrate that Padre Pio's "sores" had a natural explanation and were maintained by the application of chemicals such as iodine or carbolic acid, Bignami designed a simple procedure, which he believed would lead to a cure of the lesions.

First, any chemicals found in Padre Pio's room should be removed (the only chemical found there was iodine). Next, the doctor proposed that the friar's wounds on his hands, feet and chest were to be bandaged and securely sealed by reliable witnesses, to prevent any tampering. Each day for eight days the bandages would be changed and resealed, and the progress of the treatment was to be noted.

If the wounds were being maintained by the application of chemicals, then protecting them from external substances with bandages should cause their bleeding and size to diminish significantly. Thus, at the end of the procedure the lesions should be well on their way to being healed.

While Bignami was specifically concerned about Padre Pio's use of iodine, Padre Paolino, the Guardian of the friary, thought carbolic acid [acido fenico] had been applied to the wounds in order to stem their bleeding (14). Carbolic acid was in fact being used in the monastery at that time to sterilize the needles used for injections to prevent the Spanish Flu (15).

The Father Provincial of the Capuchins, Pietro Ischitella, agreed to the procedure. Under the precept of obedience, he ordered the monastery's Superior, Padre Paolino, and a small group of priests to implement Dr. Bignami's treatment. Padre Pietro made the friars swear that they would scrupulously follow the directives.

The Capuchin friars were extremely willing to undertake this task. Now they would be able to see for themselves the stigmata that Padre Pio was always so careful to conceal even from his brothers in religion. Padre Paolino, as the Superior, could have exempted himself, but would not let the opportunity pass. Three others were chosen to be the "reliable witnesses:" Padres Placido, Ludovico, and Basilio.

Padre Paolino later wrote: "The Father Provincial departed, and the next morning, in the presence of the witnesses, I helped to remove the habit and undershirt, together with Padre Pio's socks; along with the other Fathers, I was able to see quite clearly the mark on his chest and those on the feet and hands..." (16)

He continued, "Thus during the space of eight days every morning we removed the bandages of the preceding day after having verified that the seal was intact, and we put new ones on; and in this way we easily observed the stigmata on Padre Pio, who however suffered immensely in the depths of his heart in exposing these wounds, which he always tried to keep hidden from the eyes of others."

The Result

"Never had the wounds shed so much blood as in those days," wrote one of the witnesses, Padre Placido, in his memoirs (17). "In the morning, before he ascended the altar to celebrate Mass, we unbound the hands, and in order prevent blood from staining the vestments and altar cloths, one of us every so often dried the wounds with a cotton wad."

Each morning the seals and bandages were always found intact. On the eighth and last day so much blood issued from Padre Pio's hands during his Mass that the friars had to send for some handkerchiefs so that the Padre could dry them. Paolino wrote:

"It seemed to us that it was a very clear sign from God against the arguments of Professor Bignami." (18)

Not only was there no improvement in the condition of the lesions, but they did not even begin to heal! Instead, the bleeding on the last day was worse than before and the blood had taken on a vivid red color. Also, not only the hands, but each of the wounds bled every day, according to a signed deposition of the witnesses (19). The existence and sustained bleeding of the stigmata persisted over time, even when any possible application of chemicals was prevented. This proved very clearly that the duration and condition of the wounds did not depend on iodine, carbolic acid, or any other external substance.

Dr. Bignami had departed the friary before the experiment was over, and it is not known what his reaction was. Rev. Bernard Ruffin's well-respected biography of Padre Pio states: "He never again visited Padre Pio, although rumor had it that years later, when he was paralyzed by a stroke, he asked for Padre Pio's prayers." (20) Ruffin also noted: " . . .while the stigmata never healed, all the other wounds sustained by Padre Pio during the course of his life mended normally." (21)

Less than 2 years after Bignami's cure had failed, the Holy Office sent an Apostolic Visitor to San Giovanni Rotondo, Bishop Raffaello Carlo Rossi. He was the Vatican's official Inquisitor into the person, stigmata, phenomena and environment surrounding Padre Pio. His final report and its accompanying documentation comprised almost 200 pages, and had been kept secreted in the Vatican archives until its release in 2006. In it, Bishop Rossi stated this about the stigmata: "We can then conclude that they were not caused or preserved with physical and chemical means, which, after all, would have been in absolute contrast with Padre Pio's proven virtue . . ." (22)

Not surprisingly, the book which ignited the current carbolic acid controversy (see note 1) devotes only one paragraph to Bignami's attempt at a cure. It does not offer comments on the results and neglects to draw the obvious conclusion. Of course, doing so would have blunted the impact of its sensationalist innuendos.[5]

Other accounts of the Bignami investigation present the same picture:

Dr Amico Bignami, Professor of Pathology at the Royal University of Rome, an atheist who in 1919 had been asked by the Capuchins to examine the wounds, had speculated that they had begun as a pathological condition, aided by auto-suggestion, and had been sustained through the use of chemicals – Padre Pio had told him that he disinfected the wounds with iodine a couple of times a week or more.

Believing that the wounds would heal if Padre Pio was denied any opportunity to clean or irritate them with chemicals, Dr Bignami had all the iodine removed from the friar's cell – the friary's Guardian, Padre Paolino, was more concerned about carbolic acid – and ordered his wounds bandaged and sealed by witnesses, with these being changed and resealed each day, the treatment's progress was recorded.

Over the eight days Dr Bignami prescribed for the treatment, the wounds bled more than before, with the friars tasked with changing the bandages noting both the continued bleeding and the fact of the bandages never being tampered with. Two years later, Bishop Raffaello Carlo Rossi penned a 200-page report into the matter, observing that the stigmata "were not caused or preserved with physical and chemical means, which, after all, would have been in absolute contrast with Padre Pio's proven virtue".

Professor Luigi Romanelli, the head of Barletta's civil hospital, the first person asked by the Capuchins to investigate Padre Pio's wounds, had believed that any speculation about their origin would be beyond medical competence, but following Dr Bignami's investigation and attempted cure, he returned to the Capuchin community with Dr Giorgio Festa in October 1919. Following his own investigation, Dr Festa wrote a detailed report confidently ruling that the wounds were not due to an external trauma or the application of any chemical irritants.[6]

The originator of the carbolic acid hypothesis, we have seen, was Fr. Agostino Gemelli, the founder of one of Rome's premier hospitals (named after him).

> "Gemelli speculated that Padre Pio kept his wounds open with carbolic acid. As a result of the Gemelli assessment, the wounds were wrapped in cloth, and, according to devotees, the bleeding continued for some 50 years until they closed within hours of his death."[7]

Gemelli was an avowed enemy of Padre Pio and called him "an ignorant and self-mutilating psychopath". Gemelli was a dangerous foe in the Italy of that era because of his political and ecclesial power. As noted in a historical paper: "the term clerico-fascist was also used in the Italian political context to designate individual members of the clergy who were supporters of Fascism, like Franciscan friar Agostino Gemelli, rector of the Catholic University of Milan *and a vociferous supporter of Fascism on such issues as the invasion of Ethiopia in 1935 and the introduction of the Racial Laws in 1938.*"[8]

Padre Pio's biographer Peroni lays out the background of Gemelli's interactions with Padre Pio.

> Father Gemelli was (...) was prejudiced against Padre Pio's veracity from the beginning.

> Father Gemelli had already written much on the subject of the stigmata. He believed that all stigmata were "psychological" rather than supernatural phenomena, with the unique exception of those of St. Francis himself, and perhaps St. Catherine of Siena.

> Because of his intransigent position, Father Gemelli had some problems with the Church. He was blamed, for instance, for the slow pace of Gemma Galgani's beatification process.

The truth is that Father Gemelli never saw Padre Pio's stigmata. Padre Pio requested written authorization to show Father Gemelli his signs; he had been instructed to do so by his superiors — and he was an obedient monk. Father Gemelli had no written permission, but kept insisting, relying on his charisma and reputation. He found his equal in Padre Pio, however, who was never impressed by position or fame, and dismissed him quite brusquely.

The esteemed Franciscan Father must have felt quite insulted, and from that time on he was Padre Pio's implacable critic. Gemelli was not always truthful in respect to his meeting with Padre Pio. He claimed falsely, for instance, that he had seen the stigmata and had observed their non-supernatural origin.

Unfortunately, Father Gemelli's evaluation of Padre Pio had a negative influence on the opinion of the times. Harassment of the Capuchin friar increased to the point that he was suspended for a few years from his spiritual ministry (except for private Mass).[9]

When an eye-witness told him that he would report to Rome that he (Gemelli) had not actually examined Padre Pio's wounds, Gemelli retorted, "I'll have you destroyed."[10]

About Luzzatto's reference to the carbolic acid, Andrea Tornielli writes,

"Luzzatto raised suspicions without getting to the bottom of any of them. He cast the stone and then hid his hand. He read only parts of documents; he made huge mistakes and errors. He cited documents in which it is inferred that Padre Pio asked a pharmacist for carbolic acid and veratrine but he did not explain that on the basis of other documents, it is quite clear what Padre Pio used these things for. The "historian of the 21st century," as Luzzatto loves to call himself, never bothered to look at a 21st-century medical textbook: He would have discovered there that those acids cannot cause stigmata, nor keep them open and bloody for 50 years. Indeed, the contrary is true: They would have had a cauterizing effect."[11]

We have already seen why the charge of hysteria fails. This was clear from the earliest days as evident from this report by one of the doctors who examined Padre Pio in the early days:

> In October 1919 the Capuchin General arranged for a further medical examination, this time by Dr. Giorgio Festa, who made his first examination of the stigmatized friar in October and sent his report about a month later to the General. . . . He . . . remarked that Padre Pio seemed transformed whenever the conversation turned to spiritual topics and he was struck by the young friar's utter sincerity. He saw how little food he ate and noted his capacity, despite this fact, to devote a great many hours to hearing confessions and listening to visitors.
>
> Articles by well-known neuropathologists at that time pointed out that hysteria can produce wounds or lesions of the skin. There were those who maintained that except for St. Francis of Assisi and St. Catherine of Siena, most cases of alleged stigmatization could be accounted for in this way. Dr. Festa, who examined Padre Pio, refuted emphatically the hysteria theory. As a result of his long interviews and close questioning of Padre Pio he stated that he found complete balance and perfect harmony between the functions of his nervous system and his mental faculties. . . . Dr. Festa's final word, in view of the medical evidence, is that the origin of Padre Pio's bleeding wounds is something which we are far from being able to explain. . . . Leaving aside the scientific reports submitted by these eminent doctors, there are numerous witnesses to the genuine and persistent existence of those bleeding wounds. During his public Mass over several decades, when devout crowds pressed around the altar, the wounds in his hands were clearly visible when he raised them during the Canon. Many devout residents in San Giovanni Rotondo have frequently spoken of seeing his half-gloved hands stained with blood. We have, moreover, the reliable testimony of a number of the friars who assisted him at the altar and during the day and who were accustomed to the sight of those wounded hands. On some

occasions during his last years they also saw the wounds in his feet and side, when he required assistance in his cell."[12]

Padre Pio's stigmata were extensively examined on three occasions by respected medical professionals, Dr. Luigi Romanelli, Dr. Giorgio Festa and Dr Amico Bignami. This was done at a time when the authorities were convinced it was all a fraud. Bignami was an atheist skeptic. As we have seen, all three studies showed that the stigmata were real wounds without a natural explanation.

Notes

[1] *Who is Padre Pio?* (Rockford, Illinois: Tan Books, 1974), 9.

[2] https://www.catholicnewsagency.com/news/13863/details-of-first-investigation-into-padre-pios-stigmata-revealed

[3] Br Med J 1931; 1 doi: https://doi.org/10.1136/bmj.1.3665.581, published 04 April 1931.

[4] https://healthyliving.azcentral.com/the-effects-of-carbolic-acid-on-the-body-12362955.html

[5] http://www.sanpadrepio.com/StigmataTruth.htm

[6] https://www.irishcatholic.com/a-controversial-capuchin/

[7] https://www.nationalcatholicreporter.org/word/pfw1228.htm

[8] https://www.tandfonline.com/doi/full/10.1080/14690760701321528

[9] Andrea Monda, "And the Light Shone in the Darkness," *Inside the Vatican*, April 1999, 8-9.

[10] Peroni, L. (2002). *Padre Pio da Pietrelcina.* Borla.

[11] "The Polemics of Padre Pio," Zenit

[12] http://www.catholictradition.org/Padre-Pio/padre-pio5a.htm

Mutations – *Wrestling the Hordes of Hell, physically!*

Here is something that may come as a surprise: Padre Pio was a wrestler! Literally. Physically.

But his principal antagonist in these battles was not human. It was the Devil himself. These battles began early in his life as a priest.

> "Now, twenty-two days have passed, since Jesus allowed the devils to vent their anger on me. My Father, my whole body is bruised from the beatings that I have received to the present time by our enemies. Several times, they have even torn off my shirt so that they could strike my exposed flesh." –Letter to Padre Agostino dated 13 February, 1913 from Padre Pio[1]

> "These devils don't stop striking me, even making me fall down from the bed. They even tear off my shirt to beat me! But now they do not frighten me anymore. Jesus loves me, He often lifts me and places me back on the bed" – Letter to Padre Agostino dated 18 March, 1913 from Padre Pio[2]

Sometimes the attacks were both public and private. The Bernard Ruffin biography of Padre Pio speaks of one such attack when

> a demonically possessed woman stopped by and started hurling insults and threats at Padre Pio.
>
> As soon as he approached, accompanied by Padre Eusebio, the woman leaped at him, screaming incomprehensible words in a

deep, hoarse voice. She tried to scratch his face, but both he and Padre Eusebio were able to keep their distance. "I was terrified," Padre Eusebio wrote,"but Padre Pio less so."

"In the Name of God, go away!" Padre Pio yelled, moving past the madwoman, who didn't follow, but kept screaming, "I'll make you pay! I'll make you pay!"

The woman was taken away and a different priest began an initial exorcism of her.

Later that night the friars were awakened by a horrible noise.

A little after ten o'clock that night, the friars on the floor where Padre Pio lived heard "a frightful thud." Padre Eusebio wrote that it was "an incredible noise, as if the big door of the church was being slammed." Then they heard Padre Pio crying, "Brothers! Brothers!"

The friars rushed in to see Padre Pio on the floor, bleeding from his face. A cushion from the kneeler was underneath his head, which saved Padre Pio from suffering a major blow to the head.

The next day Padre Pio's face looked bruised and his eyes were black from the impact he suffered the night before. The woman was back in the crowd and taunted Padre Pio.

"Ah, I got my vengeance on that miserable old man!" Moreover, she kept shouting, "Ah, that cushion! That cushion!"

Later the priest resumed the exorcism and related how the woman was distraught at how that cushion saved Padre Pio.

[W]hen the priest who had been trying to exorcize her the day before began once more, the woman shrieked, "Where were you last night? I was upstairs to see the old man I hate so much because he is a source of faith. I would have done more, only the White Lady stopped me." It was said that the cushion found under Padre Pio's head was put there by the Virgin Mary.

Eventually the woman was successfully exorcised from the demon possessing her and was grateful for Padre Pio's prayers. She was liberated and became, "the most happy woman in the world."[3]

Talk about the Devil may sound fanciful or worse to the modern mind. Except for the fact that modern culture is awash with the occult, with horror and all manner of darkness.

As for the manifestation of supernatural evil, virtually everyone knows about the rite of exorcism.

A perfect case-study is an account of the phenomenon by a non-believer. William Friedkin, the agnostic director of the movie *The Exorcist*, witnessed a real-life exorcism by Fr. Gabriele Amorth, the most famous exorcist of the modern age who was the veteran of some 70,000 exorcisms.

He chronicled the experience in *Vanity Fair*[4]:

At exactly three P.M. he began to conduct the ritual of exorcism. The possessed woman, Rosa, was in her late 30s, tall and slender, with raven-black hair. …

I sat two feet away from Rosa as her torment became visible. Her family stood against a wall to my right. Father Amorth invited everyone to join him in saying the Lord's Prayer and Hail Mary. Then he invoked Saint Joseph, Padre Pio, Father Amantini, and the Blessed Virgin, asking for their protection.

Rosa's head began to nod involuntarily. Her eyes rolled back, and she fell into a deep trance. Father Amorth spoke in Latin in a loud, clear voice, using the Roman ritual of Paul V, from 1614. He asked the Lord to set her free from demonic infestation. "EXORCIZO DEO IMMUNDISSIMUS SPIRITUS." (I exorcize, O God, this unclean spirit.)

Rosa's body began to throb, and she cried out, before falling back into a trance. Father Amorth placed his right hand over her heart. "INFER TIBI LIBERA." (Set yourself free.)

She lost consciousness. "TIME SATANA INIMICI FIDEM." (Be afraid of Satan and the enemies of faith.)

Without warning, Rosa began to thrash violently. The five male helpers had all they could do to hold her down. A foam formed at her lips.

"RECEDE IN NOMINI PATRIS!" (Leave in the name of the Father.) Rosa's features slowly altered into a mask of despair, as her body continued to writhe. She was trying to rise and, clearly, to attack.

"SANCTISSIMO DOMINE MIGRA." (Let him go, O God Almighty.) Rosa did not speak or understand Latin, but she thrust forward and screamed in Father Amorth's face: "MAI!!" (Never!!)

A low buzzing sound began, like a swarm of bees, as the others in the room prayed quietly. "SPIRITO DEL SIGNORE. SPIRITO, SPIRITO SANCTO SANCTISSIMA TRINITA." (God's spirit, Holy Spirit, Holy Trinity. . . . Look after Rosa, O Lord, destroy this evil force so that Rosa might be well and do good for others. Keep evil away from her.)

Then Father Amorth called out the satanic cults, the superstition, the black magic that had possessed her. She reacted, growling, and screamed "MAAAAAAIIIIII!!!" The scream filled the room.

Another voice from deep within her shouted in his face: "DON'T TOUCH HER! DON'T EVER TOUCH HER!!" Her eyes were still closed. Father Amorth yelled, "CEDE! CEDE!" (Surrender!)

She reacted violently: "IO SONO SATANA." (I am Satan.)

The buzzing continued. Rosa grew more defiant and agitated. The room was cold, but everyone was sweating.

Except Rosa.

"RECEDE ERGO NUNC!" (Leave her now.)

"MAAAAAAAIIIIIII!"

"Answer me!"

"NO!!" "SATANA! SATANA!"

"How many demons are you?"

"Eighty legions!"

"IN NOMINE DEO QUANDO TU EXIS?" (In the name of God, when are you leaving?)

"MAAAAAAI!!!" And then, "SHE IS MINE! SHE BELONGS TO ME!"

> "She belongs to Jesus Christ!"
> "WE ARE AN ARMY!!!!"
> "Requie creature Dei" (Rest, creature of God), Father Amorth said quietly."

As remarkable as this encounter might have been, perhaps just as fascinating was the reaction of leading neurosurgeons and psychiatrists to a video of the exorcism.

> "I showed the video of Rosa's exorcism to two of the world's leading neurosurgeons and researchers in California and to a group of prominent psychiatrists in New York. ...
>
> The reaction of the neurosurgeons took me by surprise. I had expected they would quickly dismiss Rosa's symptoms as madness or unintentional fraud or suggest that she might be cured by brain surgery. They did not.
>
> They wouldn't come out and say, "Of course this woman is possessed by Satan," but they seemed baffled as to how to define her ailment, and both agreed it was not something they would attempt to cure with surgery. ...
>
> I was eager to pursue another path, one devoted to the treatment and prevention of mental disorder. I took the video to a group of some of the leading psychiatrists in the country, all in residence at Columbia University. ...
>
> LEWIS-FERNÁNDEZ: The person is expressing a pathology that is understood as possession. Our field of psychiatry can understand it as possession just on the virtue of what she's presenting, without having to take any kind of stance on whether there actually are demons, spirits. ...
>
> LIEBERMAN: ... There was no way I could explain what happened. Intellectually, I might have said it's possible, but this was an example that added credence. ...
>
> FIRST: I think all of us would agree there are things we can't explain.

I went to these doctors to try to get a rational, scientific explanation for what I had experienced. I thought they'd say, "This is some sort of psychosomatic disorder having nothing to do with possession." That's not what I came away with. Forty-five years after I directed *The Exorcist*, there's more acceptance of the possibility of possession than there was when I made the film.

The Backstory

So who or what is the Devil?

The Hebrew Bible and the New Testament testify to the reality of a realm of evil forces that existed before the material creation. Pure spirits, commonly called angels, were brought into being from nothingness by the infinite Lover. They were offered the choice of freely entering into a loving union with their Creator.

The divine Lover sought only what was best for its beloved. What was best for the beloved was to love the Source of their being unconditionally. For it is in loving the infinite plenitude of perfections that they would be fulfilled, i.e., joy-filled.

But this was not to be. The greatest of the pure spirits, Lucifer, said "No" to Love. Instantly and forever he became the ugliest and unhappiest of beings: the Satan or Adversary, the Devil. Multitudes of the spirits made the same choice and shared the same fate: eternal separation from the Source of all joy and, consequently, a state of endless misery called Hell. To be sure, not all the spirits said "No" and the greater part of them (the "good" angels) chose the path of love.

The Book of Genesis tells us how the Devil sought to derail the human experiment by luring the primordial parents of the human race into a trap – a breach of their covenant with their Creator. The consequence was catastrophe: suffering, death and impending damnation for humanity. And, for all time, the effect of the Original Sin taints human life, action and thought.

Human history cannot be understood if we do not understand the role of the occult and the diabolic. Life on earth is essentially a battle between good and evil. Underlying it all is Satan's assault on both God and humanity.

With regard to the occult, the single greatest source of God's anger in Old Testament times is the worship of idols and strange deities along with sorcery, magic and all forms of the occult. This is not because he is the "jealous Jehovah" but because all of them entail enslavement to the demonic. And that path leads only to destruction and damnation.

In the life and ministry of Christ, we witness ongoing diabolic intervention as when Satan enters Judas. We see Jesus casting demons out of the afflicted and also exchanges between the demons and Jesus. Jesus warns us frequently of Satan's interventions.

The Book of Revelation tells us the Devil and his demons have been sent here on earth. Their final destination is an everlasting abyss in which they are frozen out of contact with anyone else. But until the Final Judgment they are allowed to stay here distracting themselves from their miserable state by drawing human persons down with them. As we see in the Gospel of Luke, when Jesus drove demons out of a possessed man, he sent them into a herd of swine because 'they pleaded with him not to order them to depart to the abyss.' So they are here with us until the end of time. This is where they are 'kept for judgment.' And they are desperate. 'Woe to you, earth and sea,' says *Revelation* 12:12, 'for the Devil has come to you in great fury, for he knows he has but a short time.'

En Garde

Exorcists and certain chosen souls have directly engaged the Devil in battle to protect their fellows. Not surprisingly, this was yet another duty entrusted to Padre Pio.

> One of the first contacts that Padre Pio had with the prince of evil occurred in 1906. One night, Padre had returned to the convent of Saint Elia of Pianisi. He couldn't fall asleep that night because of the enormous summer heat. He heard the footsteps of someone

coming from a nearby room. Padre Pio thought, "Apparently, Friar Anastasio couldn't sleep either." He wanted to call out to him so that they could visit and speak for awhile. He went to the window and tried calling to his companion, but he was unable to speak. On the ledge of a nearby window, he saw a monstrous dog. Padre Pio, with terror in his voice, said, "I saw the big dog enter through the window and there was smoke coming from his mouth. I fell on the bed and I heard a voice from the dog that said, "him it is, it is him". While I was still on the bed, the animal jumped to the ledge of the window, then to the roof and disappeared."

The Devil attacked Padre Pio with numerous types of temptations. Padre Augustine also confirmed that the Devil appeared to him under many different forms. "The Devil appeared as young girls that danced naked, as a crucifix, as a young friend of the monks, as the Spiritual Father or as the Provincial Father; as Pope Pius X, a Guardian Angel, as St. Francis and as Our Lady." The Devil has also appeared in his horrible forms with an army of infernal spirits. There were other times when Padre Pio was approached by the Devil but without any apparition. He was troubled with deafening noises and covered with spittle, etc. Padre Pio succeeded in freeing himself from these attacks of the Devil by invoking the name of Jesus.

The struggle between Padre Pio and Satan became more difficult when Padre Pio freed the souls possessed by the Devil. Father Tarcisio of Cervinara said, "More than once, before leaving the body of a possessed, the Devil has shouted, "Padre Pio, you give us more trouble than St. Michael"; also, "Father Pio don't steal the bodies from us and we won't bother you."[5]

In one of the letters to Padre Agostino, dated January 18, 1912, he stated, "The Devil does not want to lose this battle. He takes on many forms. For several days now, he has appeared with his brothers who are armed with batons and pieces of iron. One of the difficulties is that they appear in many disguises. There were several times when they threw me out of my bed and dragged me out of my bedroom. I am patient, however, and I know Jesus, Our

Lady, my Guardian Angel, St. Joseph and St. Francis are always with me."[6]

Despite all the attacks, Padre Pio was never intimidated by his infernal adversary. Neither was he obsessed with the demonic. His main message to the faithful was this: If we stay away from evil, the Devil cannot harm us.

This is how he put it:

> "The devil is like a rabid dog tied up to a chain; beyond the length of the chain he cannot seize anyone. And you: keep a distance. If you approach too near, you let yourself be caught. Remember that the devil has only one door by which he enters the soul: the will. There are no secret or hidden doors."

> "The best way to avoid falling is to lean on the cross of Jesus. Never fear the tricks of satan, because even though they might be vigorous they will never move a soul who keeps itself attached to the Cross."[7]

Notes

[1]PADRE PIO DA PIETRELCINA: Epistolario I° (1910-1922) a cura di Melchiorre da Pobladura e Alessandro da Ripabottoni - Edizioni "Padre Pio da Pietrelcina" Convento S.Maria delle Grazie San Giovanni Rotondo
[2]Ibid.
[3]https://aleteia.org/2021/09/23/how-our-lady-saved-padre-pio-from-a-violent-demonic-attack/
[4]https://www.vanityfair.com/hollywood/2016/10/father-amorth-the-vatican-exorcist
[5]https://www.padrepio.catholicwebservices.com/ENGLISH/The_Devil.htm
[6]PADRE PIO DA PIETRELCINA: Epistolario I° (1910-1922), op cit.
[7]https://hbpilgrim.com/st-padre-pios-lessons-on-defeating-satan/

Radioactive – *The Healing Touch*

The canonization process whereby someone is declared a saint requires, among other things, at least two miracles that can be directly attributed to the posthumous intercession of the candidate. The two miracles supporting the canonization of Padre Pio were of a medical nature. Of course, these occurred after his death. But healing miracles were a constant when he was alive as we have seen from some of his bilocation appearances.

We have referred to the miracle involving Pope John Paul II and his friend Dr. Wanda Poltawska. Here is Dr. Poltawska's own account of the event:

> Many journalists, pursuing perhaps the sensational, have tried to convince me to give an interview. I will not dwell here on the fact that in 1962 I was to undergo surgery for what was presumed to be a tumorous growth. However, I was told just before surgery that I was healthy and could go home. I was not, at that time, conscious of anything unusual. Rather, I was ready to think that what my colleagues, the surgeons, had stated: namely that there was a five percent chance that it was merely an inflammation, and that this turned out to be the case.

> I had no idea that two letters in Latin had been written on my behalf to Padre Pio by the Archbishop of Cracow, Karol Wojtyla, the first with a request, and the second with thanks. At this time, our Archbishop was in Rome attending Vatican II. I did not know then of the existence of such a Capuchin priest, since communist Poland was completely isolated from the West, and information about Padre Pio, already well-known elsewhere in the world, had not yet penetrated into Poland, as far as I knew.

But even when I learned about the letters, I didn't want to reflect on what happened. It seemed too difficult to comprehend a supernatural intervention. A doctor tends to see natural explanations; nature herself can seemingly perform miracles and surprise doctors. It happens so often: The doctors' prognosis is positive, but the patient dies; or the opposite, the doctors expect the patient to die, but he recovers. We do not yet definitely know the power of human resistance or the forces of nature.

Therefore, in some sense I laid this question aside without trying to understand it. However, information concerning Padre Pio and his intervention in people's lives began to reach me from many sources. And then when Archbishop Karol Wojtyla returned to Cracow, I learned of the exchange of letters and who this Padre Pio is.

In 1967, I had my first opportunity to go to Rome, and in May I went to San Giovanni Rotondo, where I arrived toward evening, hoping to participate at Holy Mass the next morning. However, I was told that there were always huge crowds and that it would be almost impossible to enter and be close to the altar.

I was standing in the little square in front of the church when I saw a Capuchin priest coming from a side gate. I approached him and told him that I would love to participate in Padre Pio's Mass, but I was afraid it would be impossible, that I came from far away Poland, and that I might not be able to get another passport to make the trip in the future. He looked at me and said, "Come to this gate tomorrow at 5 a.m. and I will let you in," and that's what actually happened. The next morning, he led me through the sacristy, and I was able to sit near the altar. I could observe Padre Pio from a short distance. He was an old man at the time and walked very slowly. He celebrated Holy Mass with incredible intensity and with an expression of suffering on his face. It is impossible to find adequate words to describe this Mass. This sacrifice of the altar was truly the representation of the Passion of Christ. Even people of deep faith hardly ever can perceive this reality on the altar as a representation of Calvary. They perceive the reality of God's presence in the consecrated Host and His Blood in the chalice. But here the reality of His Passion, which is beyond our understanding, seemed to be reflected in the Holy Mass celebrated by this old man. It reflected the suffering of the Passion. Perspiration from Padre Pio's forehead ran down his face, his hands were covered with bandages, and the dark stains of the stigmata were barely visible behind his long sleeves. The agony of the man was visible.

The church, full of people, was silent, unusually silent for Italy, only interrupted now and then by a sob. The Mass lasted a long time, and when it was over, Padre Pio slowly made his way back to the sacristy with short steps. As he was passing by, I happened to be near him. He stopped for a minute, looking around at the people, then looked directly at me. I shall never forget his glance. Smiling, he came even closer to me, patted me on the head, and said," Adesso, va bene?" (Now, are you all right?) I did not answer. I had no time. What could I say?

But precisely in this moment, I knew he recognized me. In this moment I also knew that it wasn't because of a wrong diagnosis that I had found myself suddenly well several years earlier, but because this monk had come into my life in such an extraordinary way because the Archbishop of Cracow had asked for it. That is all I can say.

Many years later, when the Archbishop of Cracow had become Pope, I learned from the man who had handed the Archbishop's first letter to Padre Pio that Padre Pio said, "I cannot say no to this request."

From then on I put many difficult problems into Padre Pio's hands. I feel I am one of his spiritual children, even though I have only seen him once. He died a year later.

However, I have contact with him continually. He has done many things for people I asked him to help. They were not miracles for which I can give documentary evidence, but I know for certain that he has helped me.[1]

Today we know what was transpiring in the background at San Giovani Rotondo when the request from the future Pope was first received. Frank Rega gives us a graphic account:

> In the early 1960's Angelo Battisti held two important positions in the Church. He was the administrator for Padre Pio's hospital, the Casa Sollievo Della Sofferenza, located just across the piazza from the Capuchin friary at San Giovanni Rotondo. In addition, he worked in the offices of the Vatican Secretary of State. Shuttling back and forth between Rome and San Giovanni was a weekly occurrence for Battisti, and he was known as a close personal friend of Padre Pio's. Thus, it was not altogether unusual when in November of 1962 he was asked by a colleague in the secretariat, Guglielmo Zannoni, to deliver an urgent letter to Padre Pio. The letter had been passed on to Zannoni by a Polish monsignor who would eventually become a Cardinal, Andrej Deskur. [notes 1,2] But this important letter was not written by Deskur himself. Instead, it was composed by another Polish prelate, a bishop from Krakow by the name of Karol Wojtyla.
>
> (…)
>
> Since it was essential that the letter arrive as soon as possible, Bishop Wojtyla, acting through intermediaries, enlisted the help of Angelo Battisti in order to have it hand-delivered to Padre Pio. Battisti's positions at the Vatican Secretary of State and as the administrator for Pio's hospital, guaranteed him virtually unlimited access to the saint at almost any hour. He was told that the letter was of utmost importance, and was asked to leave at once and deliver it personally to Pio. The hastily summoned messenger later remarked: "I had never received such an urgent assignment. I quickly went home to get my car, and departed immediately." [12]

This One Cannot be Refused!

Battisti drove to the friary at San Giovanni Rotondo and headed straight for Padre Pio's room. There he found the priest seated with his head bowed over his chest, engrossed in prayer. The messenger held out the envelope, explaining that it dealt with a pressing matter. Without moving, Pio simply replied, "Open it and read it." He listened in silence as Angelo Battisti read the letter, and remained silent for some time afterwards. Battisti was now surprised that this missive had to be urgently delivered; it seemed to be similar to the torrent of grave requests about life and death matters that daily reached Padre Pio, imploring his prayers. Finally, the Padre raised his head, and with a serious demeanor turned towards the messenger. "Angelo, to this one [questo] it is not possible to say no!" Then he bowed his head as before and resumed praying.

Battisti understood that by using the term "questo", a masculine pronoun, Pio was referring to the person (this one) who sent the letter. On the drive back to Rome, he thought about the many years he had known Padre Pio, and how every single word he wrote or spoke was carefully chosen and had a profound significance. He did not use the feminine "questa," which would have referred to the request or to the letter itself. No, it was "questo" – he who sent it – that could not be refused. But who was this Polish Bishop? Though Battisti worked at the Secretariat of State, he never heard of him. Nor, he found out when he arrived at the Vatican, had any of his colleagues ever heard of Bishop Wojtyla. Yet, why had Padre Pio considered him so important? [13]

The operation to remove the tumor in Dr. Poltawska's intestine was to take place on a Friday in late November, 1962. On Saturday, Bishop Wojtyla telephoned the sick woman's husband Andrei to learn whether or not the tumor had been malignant. Andrei started to explain that the operation never took place because the doctors had found that there was nothing they could do. The Bishop immediately began to console his friend, believing that the cancer had been declared inoperable. Andrei interrupted: "Oh no, you do not understand...The doctors are confronted with a mystery... They could not find anything." [14] The growth, which had been previously confirmed as present by the doctors, had now completely disappeared! For Bishop Wojtyla, only one explanation for this cure was possible – the prayers that Padre Pio had raised to heaven.

At the time, the Poltawskas knew nothing about their friend's letter to the holy man of the Gargano, and they did not find out until later. In fact, the couple had never heard of Padre Pio, since Poland was still a closed-off Iron Curtain country, and there was little opportunity for them to learn about events in the free world. Thus, at first Wanda attributed the results to the one-in-twenty possibility that it was an inflammation which had healed on its own, and not a tumor at all. [15]

Upon hearing the good news, Bishop Wojtyla composed a second letter to Padre Pio, this time thanking him for interceding before God for this mother of four children. In the letter dated November 28, again in Latin, he clearly attributes the doctors' failure to find any diseased tissue to divine intervention.

Venerable Father, the woman living in Krakow, Poland, and mother of four children, on the twenty-first of November, prior to the surgical operation, was suddenly cured. Thanks be to God! And also to you venerable father, I offer the greatest possible gratitude in the name of the woman, of her husband, and all of her family. In Christ, Karol Wojtyla, Capitular Bishop of Krakow. [16]

Once again the bishop's letter was consigned to Angelo Battisti, with instructions from Vatican officials to immediately carry it to San Giovanni Rotondo. He departed at once, and upon reaching Our Lady of Grace Friary, the messenger approached Padre Pio in his cell. As before, Pio spoke the simple command: "Open it and read." This time Battisti himself was extremely curious, and upon reading aloud "the truly extraordinary and incredible news" he turned to Padre Pio in order to congratulate him. But the friar was immersed in prayer. "It seemed that he had not even heard my voice as I was reading the letter." [17] The minutes passed by in silence, and finally the Padre asked Angelo to keep these letters from Bishop Wojtyla, because some day they would become very important.

Returning to Rome, Battisti secured the letters in a safe place, and as the years passed, he almost completely forgot about them. Then, after sixteen years, the evening of October 16, 1978 arrived. Gathered with the crowds in front of Saint Peter's Basilica, he waited anxiously for the announcement of the name of the new pope. When he heard the words "Karol Wojtyla," Battisti was stunned. His first thoughts were of the words of Padre Pio from long ago, "Angelo, to this one it is not possible to say no!" – and then tears came to Battisti's eyes. [18][2]

Another of the well known healing miracles involved a blind woman who was born without pupils.

Among the many miracles of healing attributed to Padre Pio, some are so unusual and unique that they have been the subject of much discussion and controversy. In these particular cases, the person who has been healed lives a completely normal life afterward, even though they continue to have all the physical symptoms of their illness. From a scientific viewpoint, they are still sick. One such person is Gemma di Giorgi.

Gemma di Giorgi was born on Christmas day in 1939, in the Sicilian town of Ribera. Almost immediately, her mother realized that her eyes were different from other children's eyes. The truth was, Gemma was blind. Her mother took her to a doctor who was unable to determine the gravity of her condition. She was referred to two specialists in Palermo. They determined that Gemma had no pupils in her eyes; that nothing could be done for her blindness; that her condition was inoperable. Gemma's family was desperate, but there was nothing they could do. Her parents often took her to Mary's altar in the church to pray because they felt it would take a miracle to heal her eyes.

A relative who was a nun, advised the family to seek out Padre Pio. Her advice gave the family a ray of hope. Gemma's grandmother asked the nun to write a letter to Padre Pio on Gemma's behalf.

When the nun returned to her convent, she wrote to Padre Pio asking him to pray for Gemma. One night the nun saw him in a dream. Padre Pio asked her, "Where is this Gemma for whom so many prayers are being offered that they are almost deafening?" In her dream she introduced Gemma to Padre Pio and he made the sign of the cross on her eyes. The next day the nun received a letter from Padre Pio in which he wrote, "Dear daughter, rest assured that I will pray for Gemma. I send you my best wishes."

The nun was struck by the coincidence of the dream and the letter that followed so she wrote to the family and encouraged them to take Gemma to see Padre Pio. And so it was, that in 1947, the grandmother took 7 year old Gemma to San Giovanni Rotondo to see Padre Pio, praying and hoping all the while for a miracle.

On the trip from Sicily to San Giovanni Rotondo, Gemma's eyesight began mysteriously functioning.About halfway to their destination, Gemma began to see the sea and a steamship and she told this to her grandmother. At Gemma's words, her grandmother as well as other friends who were accompanying them, all began to pray. Nevertheless, the trip from Sicily to the monastery was very long and difficult. Gemma's grandmother was still preoccupied with the idea of seeking Padre Pio's intercession regarding Gemma's eyesight.At San Giovanni Rotondo, in the midst of a large group of people, Padre Pio singled Gemma out and called her by name. He heard her confession, and even though she made no mention of her blindness, he touched her eyes with the wounded part of his hand, tracing the sign of the cross. At the end of the confession, as he blessed her, he said, "Sii buona e santa." (Be good and saintly)

The grandmother was upset that Gemma had forgotten to ask Padre Pio for the grace of a healing while she was in the confessional and so she began to cry. Gemma became upset also and began to cry.

The grandmother went to confession to Padre Pio and in her own words, "I asked the grace for Gemma and I told Padre Pio that Gemma was weeping because, in her confession with him she had forgotten to ask this grace. I will never forget his soft and tender voice as he answered me with these words, "Do you have faith, my daughter? The child must not weep and neither must you for the child sees, and you know she sees." I understood then that Padre Pio was alluding to the sea and the ship Gemma had seen during the trip and that God had used Padre Pio to break through the darkness that covered Gemma's eyes." The same day, Padre Pio gave Gemma her first Holy Communion and again made the sign of the cross over each of her eyes.

When Gemma returned to Sicily her eyes were again examined by a specialist. The doctor, to test Gemma, held up various objects in front of her and she was able to see each one of them. She was able to count the doctor's fingers at a distance of sixteen feet. Gemma, even though without pupils, had her eyesight; she could see. The doctor declared that Gemma's eyes were in no condition to see. There was no medical explanation for it.

Many doctors from all over Italy requested to examine Gemma's eyes. This extraordinary cure, and the prophecy preceding it, aroused enormous interest in the Italian press during the summer of 1947. Gemma's sight continued to improve and she was able to go to school and learn how to read and write. She was able to lead a perfectly normal life.

Clarice Bruno, author of the book, "Roads to Padre Pio" met Gemma in May of 1967. Clarice said that despite the fact that Gemma's eyesight was functioning, she still had those foggy, strange looking eyes that are a characteristic of the blind. Clarice told Gemma that she was writing a book about Padre Pio and wanted to include in it the story of Gemma's miracle. Gemma asked Padre Pio for permission to share her story and he gave his consent. Gemma, due to the sunny and very windy weather, was wearing sun glasses on the day she made the request of Padre Pio.

He commented on this. "Why," he said as he passed his hand over her eyes,"are you wearing glasses? You see very well."

Father John Schug, author of "A Padre Pio Profile" who met Gemma and interviewed her, also testified, "She looks like a blind person. Her eyes are sallow and lusterless, but there is no doubt that she can see. I saw her reach for a phone book, check a number, and dial the number without groping."

While the doctors could not all come to a consensus on the subject of Gemma di Giorgi's medical condition, the facts that can be definitely established are these: (1) Gemma di Giorgi was born with a severe congenital defect of the eyes; (2) before the prayers of Padre Pio were enlisted, her vision was either quite defective or altogether nonexistent; and (3) afterward, though the physical structure of her eyes remained unchanged, Gemma was able to see normally–even though officially classified as legally blind.

So what must the conclusion be? Simply that while Gemma and her grandmother were traveling to San Giovanni Rotondo to ask for healing grace, the grace came to them through the intercession of Padre Pio's prayers before they had even arrived at their destination. God, for His own mysterious reasons, had wanted it to happen this way. Gemma has since traveled the world telling her story.[3]

Miracles?

Healing miracles have accompanied the propagation of the Christian Faith throughout history. But are miracles possible?

Miracles are central to the Jesus story. Those who deny the claim that Jesus performed miracles on the supposed basis of "science" are unwittingly trapped in a paradox. Modern science was possible only because of the intellectual matrix provided by the followers of Jesus. Christianity introduced a new "idea": the world was "separate" from God. For pre-Christians, as philosopher and historian of science Stanley Jaki points out, "the expression 'only begotten' had the universe for its supreme reference point." But "with Christians, inspired … [by the] only begotten Son of God, the universe could not retain its hallowed status as a 'begetting' from the 'divine,' that is, the status befitting an entity sharing divine nature."[4] The Universe was a creation of God which had a beginning and was not divine in its own right as many before believed. It followed laws embedded in it by its supremely intelligent Creator. The laws were discoverable by human beings because humans were made in the rational image of their Maker. All these novel "ideas" were part of the phenomenon of Jesus. They are precisely the ideas that freed the human mind from superstitions of a totally chaotic universe or a universe that was part of a deity and therefore not to be examined. And they are the ideas that gave birth to modern science. Hence the paradox of deploying modern science to reject the claims of its progenitor.

Also, the same Christian matrix that produced modern science took it as obvious that the Creator had a purpose in creating the Universe and would order its operations to serve this purpose. Such "ordering" sometimes meant transcending the ordinary laws of nature aka miracles.

Finally, the science we have today was made possible because Christianity enabled trust in human observation and reasoning. The skeptics did not trust their senses or their intellects. Nothing was knowable. This was a reasonable belief if the world was total chaos and humanity a reflection of that chaos. But the revelation of a Supreme Intelligence who created beings capable of reasoning and knowing and receiving Jesus' divine Life changed this view of things. Thus Christianity not only created the first universities but powered an explosion of intellectual inquiry inspired by faith in what we perceive and think. In yet another paradox, we see that accounts of miracles and the data studied by science are both dependent on human observation and the testimony of witnesses. The one difference is that miracles do not follow "laws" and cannot be "repeated." They nonetheless, like science, derive from (in most cases) the physically observable and are anchored in eyewitness testimony. And, as was the case from the start of the era of science, accounts of miracles should be evaluated using the laws of evidence proper to them.

But ever since the "Enlightenment" the idea of the miraculous was excised from acceptable discourse. The 18th century philosopher David Hume said that no amount of evidence could justify belief in events that are incompatible with the ordinary experience of humanity. 19th century New Testament critics produced lives of Jesus shorn of miracles. The 20th century theologian Rudolf Bultmann said it was impossible to believe in miracles in the age of electricity. And the achievements of technology led many to accept the faith of funda-materialism: everything (including the belief in funda-materialism!) is governed by the laws of matter.

Fortunately, things have changed considerably in recent times. The intelligentsia today finds naïve scientism embarrassing. The wise recognize that "science" is a methodology for measuring quantities. It can help us discern patterns and make predictions concerning the behavior of what is measurable. But it cannot tell us how the laws of nature originated or whether a non-physical reality can change the normal behavior of the physical. Even in our everyday experience, we see how our non-physical intentions and thoughts direct the behavior of our neurons. So why could not the infinite non-physical Mind that brought matter into being exercise new kinds of power over it when required?

Secondly, there is greater awareness of cultural arrogance and parochial thinking. Millions of people over centuries have experienced "miracles" such as scientifically inexplicable healings. Today, thousands of communities involving hundreds of millions of people, particularly in Asia, Africa and Latin America, claim to have witnessed or experienced a multitude of miracles. We cannot, like Hume, dismiss these miracles as "out of the ordinary" simply because today the claims are widespread enough to be "normal." Only cultural prejudice can lead us to unilaterally declare them superstition, fraud and self-deception. As for science, a "scientific" mind should be driven by a quest for evidence and not by a dogmatic refusal to study data that falls outside an arbitrary framework. We cannot accept all claims of miracles at their face value. Neither can we dismiss them without further ado. We should examine the available evidence and consider possible patterns that emerge from them.

This new outlook has helped re-open discussions of the miracles of Jesus. The Bultmannian dogma that the miracle stories belong to mythology is now part of the past. Scholars have begun to re-evaluate their historical basis. Craig Evans reports that the miracle stories are now both treated seriously and widely accepted by Jesus scholars as deriving from Jesus' ministry.[5] For instance, E.P. Sanders said in Jesus and Judaism that it is "almost indisputable" that "Jesus was a Galilean who preached and healed."[6] Even such radical skeptics as Robert Funk and Marcus Borg accept some of the healings and exorcisms of Jesus. Funk comments that Jesus' reputation as a healer made it probable that he did engage in healing. Borg holds that there is a historical core behind the picture of Jesus as healer and exorcist especially given that there are more healing stories about Jesus than about anyone else in Jewish tradition.[7] The fact that exorcisms were attributed not just to Jesus but to his followers suggests to scholars that Jesus did perform exorcisms. Josephus, the first century Jewish historian, wrote that Jesus was "a doer of startling deeds". As to the claims of Jesus bringing the dead back to life, it is striking that the town of Bethany was renamed as el-Azariyeh, "the place of Lazarus", by the Arabs after its destruction by the Romans in 70 A.D.

A major milestone in the modern re-discovery of the miracles of Jesus was the publication of Craig Keener's *Miracles – the Credibility of the New Testament Accounts*. This widely-acclaimed work reviews scientific and philosophical issues raised by miracles and studies reports of miracles in ancient and modern times as well as across the world.

Keener points out that every different early Christian source as well as non-Christian sources such as Josephus and Tacitus portray Jesus as a miracle worker. The Gospel accounts of miracles are a central component of the story of Jesus' life present in the earliest layers of tradition. They are eyewitness reports not later legendary stories. And these accounts are significantly different from miracle stories in Greco-Roman and Jewish works and were not therefore derived from pagan or other sources. He notes that the majority of today's Jesus scholars recognize that Jesus was experienced by his contemporaries as a miracle worker.

Keener rejects the idea that Jesus' healings were simply psychosomatic, "Among the categories of disorders that multiple attestation suggests that Jesus cured are blindness, deafness, skin disorders ("leprosy"), and occasionally death. Some summaries (e.g., in Q, Matt 11:5/Luke 7:22), not to mention specific cases, suggest that Jesus healed *multiple* cases of blindness, deafness, leprosy, inability to walk, and death. Would he have encountered so many psychosomatic cases, and primarily psychosomatic cases, of such dramatic ailments, in a one- to three-year ministry in Galilee?"[8] The same is true of the psychiatric explanation. "Some suggest that Jesus's cures of blindness, paralysis, and the like reflect his cure of a particular psychiatric disorder; yet how many psychiatrists regularly cure cases of these afflictions (especially publicly and immediately)?"

The miracles of Jesus, as recorded, are not simply capricious acts or attempts to impress its witnesses. They serve a purpose: they are "signs." They manifest Jesus' identity and his mission. They are intended to save souls, to inaugurate the Kingdom of God. In these respects, here again we have something that is unique in world history. There have been other claims of miracles in various times and places and cultures. There is no reason to rule out the possibility that some at least of these claims may be grounded in fact. But never has there been a portfolio of phenomena such as are presented in the Gospels. All of them inextricably embedded in the revelation of Jesus' identity.

Appropriately, Jesus' miracles relate to the three realms of reality: the physical world, the spiritual world and the human world (which represents the union of spirit and matter).

In the first realm we include the celestial miracles relating to the birth (star of Bethlehem) and death (solar eclipse) of Jesus, the two occasions when Jesus is at his most helpless. But there are also such miracles as the calming of the sea, walking on water, transformation of water into wine, the multiplication of loaves, securing an abundance of fish and the Transfiguration. All testify to Jesus' authority over nature.

Jesus' ministry also concerns the spiritual world. He has come to exorcize the powers of evil Demons recognize and flee from him. He casts them out of those whom they oppress. He is Lord of the spiritual world.

Finally, and principally, Jesus is concerned with the human person. He cures ailments, cleanses lepers, gives sight to the blind, raises people from the dead. "The blind regain their sight, the lame walk, lepers are cleansed, the deaf hear, the dead are raised, the poor have the good news proclaimed to them." (Luke 7:22). He is the Healer of humanity restoring body and soul with his divine power. As always, in so doing he discloses his divine identity while fulfilling the plan of God announced by the prophets ("the poor have the good news proclaimed to them.")

It is important to realize that Jesus performs his miracles in his own name. "I say to you, rise, pick up your mat, and go home." (Mark 2:11). "Moved with pity, he stretched out his hand, touched him, and said to him, "I do will it. Be made clean."" (Mark 1:41). "Little girl, I say to you, arise!" (Mark 5:41). Jean Galot asks, "Did he personally possess a power that belongs to God alone? Jesus provides the proof that the Son of Man has, on this earth, the power of God. ... [Miracles] are the sensible deployment of divine power, but they are always linked to the person of Jesus. In fact, it is in Jesus that divine power has become incarnate, and this incarnation of power is linked to the Incarnation of the person of the Son of God."[9]

Jesus told his followers that they too could do miracles in his name. "Amen, amen, I say to you, whoever believes in me will do the works that I do, and will do greater ones than these, because I am going to the Father." (*John* 14:12)

Healings

It is Jesus' promise to his followers that was the source of the miracles attributed to the intercession of Padre Pio, a point he always emphasized. "Let's leave the miracles to God, firstly because we are not capable of working them."[10] What follows is a sampling from different publications of a few more of the healing miracles associated with Padre Pio.

Padre Pio's Healings – Catherine Birri

> In 1919 in San Giovanni Rotondo, a fourteen-year-old boy experienced a miraculous healing. When he was only four years old he had caught typhus. The typhus caused the boy's back to become deformed. Once he confessed himself to Padre Pio, the priest touched him with his hands. When the boy stood up, the deformity in his back had disappeared.

Also in 1919 a gentleman from Foggia, Italy came to Padre Pio. The gentleman was sixty-two-years-old and was walking with not one, but two canes. He suffered a terrible fall from a carriage, a fall that broke both his legs. With doctors unable to help him he relied on those two canes for support. One day after confessing himself to Padre Pio, the friar told him: "Stand up and go away! You have to throw away these canes." The man carried out the order and began to walk by himself.

There was a gentleman whose left knee was swollen and in great pain for many days. The doctor told him the situation was very dire and prescribed to him a lengthy series of injections. Prior to beginning the treatment, the man went to confession to Padre Pio. He asked Padre Pio to pray for him. The man recounted, "When I was leaving San Giovanni Rotondo that afternoon, the pain disappeared. I observed my knee and noted it was not swollen anymore! Both my legs were now normal. So I immediately ran to Padre Pio in order to thank him. He said: 'You do not have to thank me, but you have to thank God!'"

In 1950, a man's mother-in-law was taken to a hospital for an operation. The mother-in-law had breast cancer. After the initial operation, it was necessary to operate on the right breast some months later. Unfortunately, due to the dissemination of the cancer cells in her body, the doctors informed her that she would not live more than four months. The man immediately went to Rotondo and waited for confession with Padre Pio. He asked Padre Pio to help his my wife's mother and heal her. The man recounted, "Padre Pio sighed long twice and said: 'We have to pray, everybody has to pray. She will recover!' It happened! In fact, my mother-in-law recovered after her operation and she went by herself to San Giovanni Rotondo to thank Padre Pio who smiling told her: 'Go in peace, my daughter! Go in peace!'" Instead of four months, his mother-in-law lived for nineteen more years.

A woman said, "In 1947 I was thirty-eight years old and I had been suffering cancer of the intestine. The cancer was found by x-ray, and it was decided to operate. Before going to the hospital I wanted to visit San Giovanni Rotondo to meet Padre Pio. My husband, my daughter and one of her friends, took me to him. I desired to confess my sins to Padre Pio and to speak with him concerning my disease, but it was not possible to meet him. My husband told my problem to one of the friars. That friar was moved and he promised to report all to Padre Pio. In a short time, I was asked to go into the corridor of the convent where Padre Pio would pass. Padre Pio walked through the crowd, but he was interested only in me. He asked me the reason for which I was so distraught and told me I was correct about the surgeon. After that, he encouraged me and he said he'd pray to God for me. I was amazed. In fact, he did not know the surgeon who was going to perform the operation, and nobody told him that I was the right person to speak to in the crowd. Yet, he appeared to know both the surgeon and me. I faced my operation with hope and with peace. The surgeon was the first one who spoke of a miracle. He performed an appendectomy despite previous x-rays that showed cancer. The surgeon did not believe in God, but since then he has put a crucifix in each room of the hospital. There was no evidence of cancer anymore. After this miracle and a short time of rest I returned to San Giovanni Rotondo in order to meet with Padre Pio. The saintly friar was going to the sacristy when he suddenly stopped and smiled at me - he said: 'What do you know? You have come back here!'...and he gave me his hand to kiss, which I affectionately held in my hands."

One of Padre Pio's miracles occurred in 1953, to a lady with abdomen problems. She had a medical check-up and x-rays because of pains in the abdomen she was experiencing. After some investigation, she learned that an immediate operation was necessary. A friend of the lady, who knew of her problem, suggested to her: "write a letter to Padre Pio in order to ask for his prayers and help". The woman received an answer in which Padre Pio suggested that she go to the hospital—he would pray for her there. She went to the hospital and again had a medical check-up and x-rays just before the operation. The same doctors were astonished to find she no longer had the serious illness anymore. After forty years the woman is still grateful to Padre Pio for his help.

A year later, in 1954, a railwayman fell ill with a strange disease, which immobilized his legs. At the time he was forty-seven-years-old. Without success many different doctors treated him. After two years of treatment, the man was going to have to retire from his work. As the situation got worse, his brother suggested to him to travel to San Giovanni Rotondo, to see the friar there to whom God had given gifts. The man went to San Giovanni Rotondo with his brother's assistance. In the Church, he met with Padre Pio who said, "Let that railway-man pass!" Remarkably, Padre Pio had never met the man before and he knew he was a railwayman. Padre Pio and the man spoke to each other for a few minutes. Later, Padre Pio put his hand on the man's shoulder. He consoled and encouraged him with a smile. As soon as the man left Padre Pio, he realized he had been healed. He no longer needed tools to help him walk; now he could walk on his own.

A priest brought his friend, a Jewish man named Lello Pegna, to visit Padre Pio in 1919. Pegna was completely blind in both eyes. The priest had brought him to Padre Pio in the hopes of his friend being cured. Padre Pio told Pegna, "The Lord will not grant you the grace of physical sight unless you first receive sight for your soul. After you are baptized, then the Lord will give you your sight." A few months later Pegna returned baptized despite the protests of his family. In the following months his sight was fully restored. For the next thirty years, his sight remained perfect. (…)

St. Padre Pio's faith and confidence in God's healing power was unparalleled. He shows us all that the power of prayer can yield beautiful, miraculous results. He was a conduit of God's grace, love, and mercy. [11]

Angelo Cantarelli

Angelo Cantarelli from Parma used to make trips by train to San Giovanni Rotondo to bring offerings for Casa Sollievo. Once, in the way back he had a stroke, and the train made an emergency stop, and he was recovered in the clinica Villa dei Pini, near Civitanova Marche. He was in a coma, and Padre Pio sent Enzo

Bertani, from the Board of Casa Sollievo, to assess the situation. Bertani: "Padre, you must obtain the grace of fully recovery for him. In Parma they are saying that he went to take offerings to Padre Pio, and got a thrombosis." Padre Pio; "Tell the family to be calm. I have asked the Lord to let him have ten more years of life." Angelo recovered, returned home, and was fully involved in his business. He died ten years later.[12]

Nicolino

When dr. Nicola Bellantuono was still a medical student he asked Maria Ciaccia, who was going to see Padre Pio, to tell him about a difficult exam he was about to take. Padre Pio: "It will go all right." Then, gazing into space, he added: "Poor boy, poor boy." Few days later Nicola started having severe abdominal pains. His father who was a doctor and three of his colleagues, diagnosed acute peritonitis and said there was nothing to be done about it. Maria went to Padre Pio again, and said: "Padre, Nicolino is dying." Padre Pio: "What do you mean? Everything will go all right." Nicolino fully recovered and became a doctor in Casa Sollievo.[12]

Notes

[1]*Lay Witness*, October 1999

[2]http://www.sanpadrepio.com/poltawska.htm

[3]https://padrepiodevotions.org/pray-hope-and-dont-worry-issue-11-april-june-2002/

[4]Stanley Jaki, *The Savior of Science* (Washington D.C.: Regnery Gateway, 1988), 70-1.

[5]Craig A. Evans, "Life-of-Jesus Research and the Eclipse of Mythology", *Theological Studies*, 54, 1993, 19.

[6]E.P. Sanders, *Jesus and Judaism* (London: SCM, 1985), 11.

[7]Marcus J. Borg and N.T. Wright, *The Meaning of Jesus: Two Visions* (San Francisco: HarperCollins), 66.

[8]Craig Keener, *Miracles – the Credibility of the New Testament Accounts (Ada, MI:* Baker Academic, 2011), 633.

[9]Jean Galot, *Who is Christ? A Theology of Incarnation* (Chicago: Franciscan Herald Press, 1981), 156-6.

[10]Padre Pellegrino Funicelli, *Padre Pio's Jack of All Trades* (San Giovanni Rotondo, Italy: Our Lady of Grace Capuchin Friary, 1991), 134.
[11]https://www.coraevans.com/blog/article/the-most-unbelievable-miracles-of-st.-padre-
[12]https://caccioppoli.com/St.%20Padre%20Pio%20Prophecy.html

Search Engine – *Reading Minds*

Another distinctive dimension of the Padre Pio repertoire was his charism of "reading souls." This was most evident in the confessional where penitents would be reminded of forgotten sins. But it was also on display in his everyday encounters. Consider the accounts of these "celebrity" confessions:

> The actor Carlo Campanini went to confession to Padre Pio in 1950. Padre Pio said, *"Begin in 1936."* *"But I confessed few days ago."* *"I told you to begin in 1936."* Campanini remembered what he had done in 1936 and had been ashamed to confess. He concluded: *"That confession changed my life, and I haven't missed daily Mass since then."*[1]

> Dr. Ezio Saltamerenda convinced sculptor Francesco Messina in 1949 to visit Padre Pio with him. Padre Pio asked if he wanted to confess. He said, *"maybe but I'm not prepared."* Padre Pio: *"Don't say anything to me. Just answer."* *"Then he began to list my sins with incredible precision. It was as though he could read my soul."*[2]

The charism of reading souls was not, of course, unique to Padre Pio. St. John Vianney, the Cure of Ars, another of the Church's great confessors, was renowned for this. Others with the charism included St Anthony of Padua, St John Bosco, St Philip Neri, St Francis of Paola, St Joseph of Cupertino and St Paul of the Cross to name just a few.

How was this be possible? It should be remembered that all our everyday communications with each other is soul-to-soul. It is not the neurons that drive our thoughts but our thoughts that drive our neurons. Although, in our current state, we use the physical as a vehicle of the mental, there is a distinction between the two. At the moment of death, there is a separation of soul and body and any communication at that stage would have to be without the body.

This is why the intercession of the saints is possible. Until the universal resurrection of the dead, the saints in Heaven (with the exception of the Virgin Mary) are separated souls. Nevertheless we can communicate with them and they can intervene in earthly matters. Such intervention is apparent in the Book of Revelation – "I saw underneath the altar the souls of those who had been slaughtered because of the witness they bore to the word of God. They cried out in a loud voice, 'How long will it be, holy and true master, before you sit in judgment and avenge our blood on the inhabitants of the earth?' Each of them was given a white robe, and they were told to be patient a little while longer until the number was filled of their fellow servants and brothers who were going to be killed as they had been." (*Revelation* 6:9-11).

The souls of Padre Pio and the other saints – while still on earth – were endowed by God with the gift of "reading souls" without a physical intermediary. This is because all their human interactions were directed to the salvation of those around them.

Below are excerpts from reports of Padre Pio's reading of souls including some that are positively amusing

An Assortment

Angelina Serritelli from USA came to Padre Pio for Confession and First Communion knowing only English. *"I confessed and we understood each other. He spoke in English."*

Gaetana Caccioppoli was told that today was her turn to confess with Padre Pio, but in reality her turn was a week away. On the church's square she said, *"If you are Padre Pio send your friar right away to call me."* Not even 5 minutes had passes that a friar came to her from the church: *"Are you signora Caccioppoli? Padre Pio is waiting for you. He says that it's your turn to confess."*

Francesco Cavicchi and his wife visited Padre Pio in June 1967. He had confessed three days before, but wanted to confess to Padre Pio anyway. The rule was at least seven days. He stood in line and when his turn was approaching, he got agitated. But Padre Pio calling him from the line said: "Come forward, my son, I have been waiting for you for a long time." He started the confession asking: "How many days has it been since last confession?" Francesco said that he couldn't remember. Padre Pio: "You have a short memory, don't you. But let me ask you this. Do you remember the bombardment In Rimini many yeas back? Do you remember the air raid shelter? Do you remember the trolley bus? But why I'm asking you to go back in time? You cannot even remember what you did less than a week ago!" At that point, Francesco started recollecting that in November 1943, when he was 28 years old, he was riding the trolley bus with about ten other people, including a middle aged monk. Then the bombs started falling. Francesco had difficulty getting off the bombed bus to reach the air-raid shelter and thought he was about to die. The monk helped him. The capuchin monk once in the shelter began to recite the rosary and inspired calm and confidence in everyone. After the sirens gave the 'all clear" signal, the capuchin monk was the first to leave. Suddenly Francesco: "Were you the monk?" "Well, who do you think it was?"

"Father my daughter is sick." "And you are much sicker than your daughter." "No, no, I'm feeling very well." "How can you be well if you have so many sins on your conscience? I see at least 32 of them."

A woman was on her knees hoping to kiss Padre Pio's hand when he passed. He passed but didn't stop and went on. She felt bad, and in her heart complained. Padre Pio turned around and came straight to her: "OK. Here is my hand."[3]

Coward

A spiritual son of Padre Pio who lived in Rome omitted out of shame to make his customary small reverence when he passed near to a church in the company of some friends. Suddenly he heard Padre Pio's voice saying: "Coward!" After a few days he went to San Giovanni Rotondo, where he was reproached by Padre Pio: "Be careful! This time I have only scolded you; next time I will give you a slap."[4]

The Smoker

A gentleman said, "I had decided to give up smoking and to offer this small sacrifice to Padre Pio. Since then, every evening, with the intact packet of cigarettes in my hand, I stayed in front of his image telling him, 'Father...it is one...' The second day, 'Padre, it is two...' Three months later, I went to San Giovanni Rotondo to see Padre Pio after doing the same thing every evening. 'Padre,' I told him as soon as I saw him, 'I haven't been smoking for 81 days, 81 packets...' Padre Pio said, 'I know everything you know, you made me count the packets every evening.'[4]

The Englishwoman

A lady came from England to have her confession heard by Padre Pio. She went to his confessional but Padre Pio closed the window saying: "I am not available to you." The woman stayed for several weeks and during this time, daily returned to his confessional and daily was turned away. Finally, Padre Pio consented to hear her confession. She asked the Padre why he made her wait so long to be heard. Padre Pio answered: "And You?" "How long have you made Our Lord wait?" You should wonder how Jesus could welcome you after you committed so many sacrileges. You have delayed your judgment for years, besides sinning against your husband and your mother you have received Holy Communion in mortal sin. The woman was stunned and reformed. She cried when she received absolution. She returned to England a few days later, very happy.[4]

Benedict XV

A bishop told the Pope: "Padre Pio is a cheat." Benedict XV: "You might be badly informed. Why don't you visit him." The bishop obeyed the pope and went unannounced. At the train station of Foggia two friars approached him. "Padre Pio sent us to accompany you to San Giovanni Rotondo." The bishop was flabbergasted. He hadn't told anybody about the trip. He told the friars: "But this is impossible! No one knew of this trip, except His Holiness and me. I have learned what I needed to know." He took the first train back to Rome.[5]

Enzo Picciafuoco

Enzo Picciafuoco from Campobasso testified that he had been told that he could ask Padre Pio for something by thinking about it during Mass. He did. Later, in the hallway, among the crowd of faithful he told Padre Pio aloud from afar: "Father, pray for my sister." Padre Pio turning in his direction said: "You already told me."[5]

Notes

[1]https://sacredheartlombard.org/st-pio-of-pietrelcina
[2]https://caccioppoli.com/Close%20encounters%20with%20Padre%20Pio%20in%20the%20Confessional.%20Baptisms,%20Communions,%20Weddings.html
[3]https://caccioppoli.com/Padre%20Pio%20facts.html
[4]https://www.padrepio.catholicwebservices.com/ENGLISH/knowledge.htm
[5]https://caccioppoli.com/St.%20Padre%20Pio%20Prophecy.html

Multi-media –
The Scent of Heaven

"But thanks be to God, who always leads us in triumph in Christ and manifests through us the odor of the knowledge of him in every place. For we are the aroma of Christ for God among those who are being saved and among those who are perishing, to the latter an odor of death that leads to death, to the former an odor of life that leads to life." 2 Corinthians 2:14-16

The charisms of Padre Pio were gifts of God. He did not ask or expect them. They were just handed to him as he went about his mission of "saving souls." Among these was the gift traditionally called the "odor of sanctity." It is the gift of a fine aroma associated with a saint's person or clothing or surroundings or presence.

Padre Pio was perhaps the most aromatic of all the saints. Many of those who saw or met or were associated with in some way encountered the aroma of roses or violets or other fine perfumes as well as incense. His wounds, in particular, gave off a pleasant aroma. But people also "smelt" him when they sought his intercession.

Fr. Charles Carty notes that "The perfume was not constant. The opinion of those close to Padre Pio is that whenever anyone noticed the perfume it was a sign that God bestowed some grace through the intercession of Padre Pio."[1] Fr. Carty chronicled many cases of Padre Pio's "odor of sanctity" in his book on Padre Pio.

Dr. Festa, one of the first medical professionals to examine the stigmata, back in 1919, stated:

"On my first visit I took from Padre Pio a small cloth stained with blood, do a microscopic examination back in Rome. I am entirely deprived of the sense of smell, but the persons that were with me in the car smelled a fragrance very distinctly, and said that it corresponded to the perfume emanating from Padre Pio. They didn't know that I had the cloth, enclosed in a case."[2]

Bishop Carlo Rossi carried out a canonical investigation in 1921. Bishop Rossi "was particularly impressed by the perfume, and examined the cell and found only plain soap. "This very intense and pleasant fragrance, similar to the scent of violet, I have smelled it. It is sensed in waves, when he walks by, in his spot in the choir, even from a distance. The scent remained with hair of Padre Pio's that had been cut two years before, as it also attached itself to the stigmatic blood and bandages."[2]

Said one of his fellow Capuchins, Padre Romolo: "In the small drawer of the nightstand in my cell there was a bandage from Padre Pio's side. This small drawer has been in my cell for a year now, and every time I open it I notice Padre Pio's scent."[2]

There are many other accounts of the fragrance of Padre Pio, a few of which are excerpted below:

Canada

> A gentleman from Toronto said: "In 1947, my wife that fell seriously ill and was hospitalised in Rome, requiring surgery. I travelled to St. Giovanni, Rotondo to see Padre Pio. While I was there, the good Padre heard my confession. After receiving the absolution, I talked to the Padre about my wife's condition. Then I asked him: "Father, help me to pray!" Immediately I smelled a delicious and persistent perfume that surprised me. I returned home in the late evening. As soon as I opened the door, I smelled again that same scent that I had smelled when I was next to Father Pio. I became confident. My wife underwent the operation without any problems. I told her the marvellous experience that I had and together we thanked Padre Pio."[3]

England and Switzerland

A married couple in England were experiencing problems and were at a point of despair. They did not know where to turn to. Someone spoke to them about Padre Pio. They wrote to Padre Pio about their problems but did not receive any answer. They, then decided to go to St. Giovanni Rotondo to meet the Padre and obtain his wise counsel. From England to Puglia (Italy), the journey was long! Moreover, it was in the middle of winter with snowy conditions prevailing. Despite the weather, they embarked on their journey. They spent their first night in Bern in a lowly hotel for that was all they could afford. Doubts filled them once more. They wondered what they would do if the Padre did not receive them. Should they continue with their journey or turn back. As they were debating on their options, the room was filled with a sweet, intoxicating perfume, that had a very relaxing effect on them. The woman began to look for the source of the perfume and thought that some distracted traveller, before them had left a bottle of perfume in the room. The search for the perfume's source was fruitless! Awhile later the perfume smell faded away and the room emanated the usual odor of a stench. The married couple shared their experience with the innkeeper. He was surprised as it was the first time that any of his clients believed to have smelled perfume in their room. The couple instead took it as a sign to continue with their journey to St. Giovanni Rotondo and to meet Padre Pio. When they got there, the young man, who was well versed in Italian, spoke to the Padre in Italian: "We wrote to you, Father, but since you have not answered us…" "What?" Padre Pio interrupted saying: "Why do you tell me I did not answer you? Did you not smell anything that evening in the Swiss hotel?" The two lovers were joyful and full of thankfulness. They understood then, the perfume that they had smelled in the room of the hotel, was the perfume of Padre Pio. With just a few words of counselling, Padre Pio helped them to resolve their problems."[3]

Bologna

A lady from Bologna, who was 24 years old, had her right arm fractured. The same arm had been operated three years prior because of a serious accident. After an operation followed by a long painful recovery the surgeon told the girl's father that she would not have use of the arm anymore. In fact the arm was completely immobile, because of the removal of a section of the shoulder blade. A bone graft had not succeeded! Father and injured daughter departed for St. Giovanni Rotondo. Padre Pio met them, he blessed them and he declared: "Above all do not despair! Confide in God! The arm will recover." Toward the end of July, 1930 the woman returned to Bologna without any improvement of her arm. Was it possible to think that Padre Pio had been wrong? For months nobody thought anymore of the problem. On the 17th. of September (the celebration of the stigmata of St. Francis), suddenly, the apartment where the family lived became scented by a delicious odor of jonquils and roses. The scented phenomenon lasted about a quarter of an hour while everybody tried to understand where that perfume originated. From that day the girl started to use her arm again. An X-Ray which she preserved jealously, showed that the bone and the cartilage were completely fused and healed.[3]

The Distracted Lawyer

A lawyer who was very devoted to Father Pio says: "Once I was in the old church in the monastery attending one of Padre Pio's long Masses. At the moment of the consecration of the Sacred Host, I became distracted. I was the only person standing in the middle of the crowd of believers who were all kneeling. Suddenly, I was overwhelmed by a powerful aroma of violets. The scent was so strong it jolted me back to the present moment. Looking around me, I knelt down without thinking anything about the unusual perfume that hovered above me. As was my custom, after the Mass was over, I went to greet Padre Pio. He welcomed me saying 'Were you a little disoriented today?' I sheepishly said 'Yes, Padre. I have been a bit absent-minded today, but fortunately, your perfume woke me up.' He said simply.: 'For you, perfume is not necessary. For you, slaps are necessary.'"³

Padre Pio was aware of the fragrance associated with him but could not explain it:

> Friar Modestino said: "Once I was on vacation at St. Giovanni Rotondo. I went to Padre Pio in the sacristy that morning to serve the Holy Mass, but other monks were already arguing in the sacristy as to who would have this privilege. Padre Pio interrupted those discussions by saying – 'only he will serve the Holy Mass' - and he pointed me out! I accompanied Padre Pio to St. Francis altar, I closed the gate, and I started to serve the Holy Mass in profound awe. When the Mass got to the point of the "Sanctus" I suddenly had a desire to smell again that indescribable perfume that I had already perceived when I had kissed Padre Pio's hand. The desire was immediately granted me and it was like I was wooed by St. Pio's perfume. The perfume increased more and more, so much so that the perfume caused me to breathe irregularly. I leaned my hand on the communion rail so as not to fall! I was about to faint when I mentally asked Padre Pio to save me from embarrassment in front of the people. In that precise instant the perfume disappeared. In the evening, while I accompanied him to his room, I asked Padre Pio for an explanation about that phenomenon. He answered me: "My child, I am not able to explain it. God intervenes to allow somebody to smell the perfume whenever He wants."³

Padre Pio's presence after his death is also marked by fine aromas.

> Another person once remarked : "Some years ago, I had a heart attack. The doctors recommended I undergo surgery to improve my condition. I entered the hospital at once. It was in June 1991. During the surgery, the doctors did a quadruple bypass. When I awoke from the anesthesia, my leg and right arm were paralyzed. I was really discouraged. After awhile, my faith returned and I began to pray to Padre Pio for help. I pleaded with him for three days. On the third day, as I was finishing my prayer, despite being

surrounded by other patients, I noticed something significant. I caught a whiff of an overwhelming aroma of flowers. When the perfumed aroma faded, I felt a sensation in my right leg. I knew at once my prayers had been answered."[3]

Notes

[1]Charles Mortimer Carty, *Padre Pio – The Stigmatist* (1963 by Radio Replies Press, St. Paul, Minnesota. 1973 by TAN Books), https://archive.org/stream/PadrePioTheStigmatistCartyCharlesMortimer6228_201810/Padre%20Pio%20the%20Stigmatist%20-%20Carty%2C%20Charles%20Mortimer_6228_djvu.txt
[2]https://caccioppoli.com/St.%20Padre%20Pio%20Perfume.html
[3]https://www.padrepio.catholicwebservices.com/ENGLISH/Perf.htm

Genetic Re-engineering –
"Father-ing" Children Worldwide

"Even if you should have countless guides to Christ, yet you do not have many fathers, for I became your father in Christ Jesus through the gospel." 1 Corinthians 4:15.

You are witnesses, and so is God, how devoutly and justly and blamelessly we behaved toward you believers. As you know, we treated each one of you as a father treats his children, exhorting and encouraging you and insisting that you conduct yourselves as worthy of the God who calls you into his kingdom and glory." 1 Thessalonians 2:10–12.

"I have made a pact with the Lord: I will take my place at the gate to paradise, but I shall not enter until I have seen the last of my spiritual children enter." Padre Pio of Pietrelcina

Padre Pio had all the charisms of all the saints. But he went beyond even these charisms with a dimension that is unique to him: he became a spiritual father to all those who asked it of him and promised that he would not enter paradise until all his children had "gotten in." There was one condition he placed: do not let me lose face by going astray.

Christians know that the goal of life is to become children of the Heavenly Father by becoming adopted brothers and sisters of the incarnate Son through the agency of the Holy Spirit ("As proof that you are children, God sent the Spirit of his Son into our hearts, crying out, 'Abba, Father!'" *Galatians* 4:6). By becoming children of the Father we become his heirs and can enter the eternal Promised Land prepared for us before the foundation of the world.

But for this to happen we have to choose to become a child of the Father and remain faithful to this choice for the rest of our lives. This is where all of us need help. And this is why Jesus gave us his Church, his Mother and St. Joseph, his virginal father, and his apostles and saints. They are here to help us remain on the straight and narrow path that leads to our eternal destiny.

We see how St. Paul was a spiritual father to the Thessalonians and the Corinthians. Likewise, the Holy Ones of Christendom have served as spiritual guides and directors to the faithful of their times. Padre Pio is one such spiritual guide and father – but one who, from the abundance of his love, has promised to stay with his children until they enter eternity.

At a human level, fathers feel a particular responsibility for looking after their children and ensuring their well-being no matter what their age. Some coaches of sports teams say that once they hire a player they feel a responsibility for the welfare of that player for the rest of their lives.

When it came to his spiritual children, like St. Paul, Padre Pio was "exhorting and encouraging you and insisting that you conduct yourselves as worthy of the God who calls you into his kingdom and glory."

It was in 1910 that Padre Pio, with the permission of his superiors, took on the responsibility of becoming a spiritual father to all who asked. The book *Padre Pio at the Threshold of Paradise* by the Italian author Saverio Gaeta studies Padre Pio's promise to his spiritual children. Andrea Tornielli gives an overview of the material in the book:

> Padre Pierino Galeoni, a priest who was a spiritual child of Padre Pio [said]: "He revealed to me that he asked Jesus to be able to be not only a perfect victim, but also a perennial victim, that is to continue to remain a victim in his children (…) up until the end of the world. He confirmed to me that he had obtained from the Lord the mission of being victim and father of victims until the last day".
>
> Thus is explained the reason behind that oft-repeated phrase, and the profound significance of that promise, of his remaining "on the threshold" of Paradise in order to draw within the very last one of

his spiritual children. When he beheld the good intentions of a penitent, Padre Pio allowed himself to offer this type of assurance. Little Anna Tortora, at the day of her Confirmation, told him she wished for a gift. The saint asked her: "What do you desire? A doll, a little book?", and the child replied: "I wish, Padre, that when you go into heaven, make sure that there is a place there for me too." "Are you certain I will go there?" answered Padre Pio; and the child: "And if you don't get there Padre, who else can get there?" At that point Padre Pio conceded: "Good, I promise you that, if I go there, I will carry you up too."

(…) There were many of his spiritual children who asked of the Padre, especially when was approaching the time of his death: "Now that you may leave us, what are we going to do without you?" He replied, gruffly but at the same time playfully, "Silly person, I will be here in your midst, more than before. Come visit my tomb. Before, in order to speak to me, you had to wait. Then, it is I who will be waiting there. Come to my tomb and you will receive more than you did before!".[1]

Padre Pio's promise of spiritual fatherhood was extended to people who had never met him. And it went beyond his earthly existence. Tornielli continues,

But the great and most generous embrace of the friar is extended also to those who were not able to meet and get to know him, those who do not have the opportunity of visiting San Giovanni Rotondo, and those many devotees that are not able to pray at his tomb. Thus reveals brother Modestino of Pietrelcina, the compatriot of Padre Pio that has come to be viewed as his spiritual successor. Already many years prior, while the saint was still alive, he pondered over this problem. "I was meditating," he recounts, "on the benefits that were gained by those who are accepted by the Padre as his spiritual children. Then I thought with regret about all those that are not able to go to San Giovanni Rotondo to ask of Padre Pio to be adopted spiritually, and of those, even less fortunate, who would not learn of the Padre until after he had died".

One day while at confession with Padre Pio, an inspiration was taking form. "I would like to take on, as the Padre's spiritual children, all those who commit to reciting, each day, one Rosary, and have a Mass celebrated from time to time, according to his intentions. Can I do this, yes or no?" Padre Pio, spreading out his arms, looked up to heaven and exclaimed: "And how could I deny such a great benefice? Do what you ask of me and I will assist you." And later he gave new encouragement: "My son, enlarge as much as you can the numbers because they are more favored before God than I am. Relate to them that I give them all of my soul, in order that they may persevere in prayer and in doing good."

In these words is confirmation of the promise. For this purpose, each evening from 8:30 to 9:00 the immense spiritual family of Padre Pio gathers, ideally, in the crypt at San Giovanni Rotondo, around the tomb of the saint of the Gargano, in order to recite the Rosary. Anyone in any corner of the Earth, can benefit from the extraordinary "insurance of eternal life" of Padre Pio. Whoever wishes, at any time, can become a spiritual child of Padre Pio of Pietrelcina, simply uniting with devotion to this recitation, also occasionally having a Mass celebrated according to the intentions of the holy Capuchin, and pledging to repudiate sin.[1]

Padre Pio's Statements on His Spiritual Fatherhood

His spiritual paternity was an important part of Padre Pio's mission and he spoke of it often in response to questions from his spiritual children[2]:

"When the Lord entrusts a soul to me, I place it on my shoulder and never let it go."

"I accept you as spiritual child; and don't make me lose face."

"I see all my children who come to the altar, as if in a mirror."

"I love my Spiritual Children as much as my own soul and even more."

"Once I take a soul on, I also take on their entire family as my spiritual children."

"I tell Mary: Here are the children of your Son."

"Woe to those who lay a hand on my spiritual children."

"To my Spiritual Children: my prayers for you will never be lacking"

"If one of my spiritual children ever goes astray, I shall leave my flock and seek him out."

Padre Pio saw his spiritual paternity continuing after his earthly departure[3]:

"My true mission will begin after my death."

"When I am in heaven I will be able to do more."

"I belong entirely to everyone."

"Everyone can say: Padre Pio is mine."

Padre Pio's Continuing Intercession for His Spiritual Children

Each month, those who became spiritual children of Padre Pio long after his death write to the monks at San Giovanni Rotondo testifying to his intercession. Here are a few instances from *The Voice of Padre Pio*, the official publication of Padre Pio's Our Lady of Grace Capuchin Friary in San Giovanni Rotondo, Italy:

Major cancer: I was the recipient of what my husband and I consider a miracle from Padre Pio. I was told I would be opened for major cancer, since I was also a nurse at the hospital I knew the score. The surgeon told me very bluntly, "We are opening you up for major cancer, with the expectation of finding it, and praying we don't." The night before surgery my husband awakened about 2 a.m. to the scent of roses (a miracle in itself, since he was out of the hospital from nose surgery and could smell nothing at all)… Suddenly he heard a voice (no apparition) and it said, "Rudolph, don't worry they won't find cancer, she'll be alright." My husband was stunned. The next morning I was in surgery six hours and my doctor came into my room smiling and saying, "Do you know you had a miracle? Ther3 isn't a trace of cancer in your body anywhere. A large tumor was removed and it was benign." I made a thanksgiving pilgrimage to San Giovanni Rotondo.

On Vacation: I want to thank Padre Pio for helping me. In March 1988, I was on vacation in Atlanta, USA. Someone broke into my hotel room and took everything, including passport, credit card, cash, and the return airline ticket home via London. I had nothing left. I contacted the police. I was hysterical but cried and prayed to Padre Pio for help. I did not want to inform my mother who was sick. The police were nice, but I would have to phone Ireland and get cash for the new airline ticket. It would take four days and my parents would have to be informed. I was originally to leave for home the following evening, and due back to work two days later. That night I prayed to Padre Pio. I knew he would help some way. The following day I called to Traveller's Aid. While in their office the police called. They had recovered the airline tickets, passport, and credit card. Only the cash was missing. The Officer said that it was a million to one chance that they were recovered in a city like Atlanta. It would be a miracle as usually it would take months… Also, as I searched a coat pocket I came upon 70 dollars, which I had forgotten about, but put there earlier to pay for a taxi. I changed this to sterling and returned home 5 pounds of it left over.

Accident: One night I forgot to unplug my electric heater in my bedroom. Right away before going to sleep I got the wonderful scent of roses. I got out of bed to see what I had done wrong and soon found out I had forgotten to unplug the heater. I said thank you Father Pio so much. The scent of roses disappeared.

Child Cured: I am writing to you about our daughter Sharlene. Sharlene was born on 8th September 1986 with a hole in her heart and was attending the hospital regularly. The hold was in the bottom part of the heart and the chances of curing on its own were very poor. In December 1989 we were told by the doctor that they would probably have to do surgery when she would be five years old. This was very disturbing news for us. Then I heard of a woman who had a Padre Pio relic and I arranged to meet with her. I met this woman in January 1990 and saw her on two occasions when the relic was placed on Sharlene. I was told by her that Sharlene would not have to go through surgery. After this I said the novena to Padre Pio each night. When I took Sharlene back to the hospital in December 1990 the doctor examined her and after ECG test the new was very good. The hole in her heart was completely closed. This was all with the help of Padre Pio.

Mobile: I thank and praise God for having granted my request through the intercession of Padre Pio. After three years in a wheelchair, walker, cane, etc. and being told by doctors that nothing could be done, I am now mobile. I did not ask for a cure but to be healed enough to enable me to attend daily Mass. Through Padre Pio's intercession, my request was granted and more… Padre Pio has become a friend. He spoils his children as my appeals for several minor favours have always been answered. I thank the Lord for working his marvels through His holy servant, Padre Pio.

How Do You Become a Spiritual Child of Padre Pio?

Those who wish to become spiritual children of Padre Pio can write to his monastery asking to become his spiritual child. The address for the monastery is:

Convento Frati Minori Cappuccini
Santa Maria delle Grazie
71013 San Giovanni Rotondo (FG)
ITALY
Email address – cappuccini@conventopadrepio.com
Website – www.conventosantuariopadrepio.com
Tel – +39 0882 4171/417600

Alternately they can write to any Padre Pio center affiliated to this monastery in their own country.

Spiritual children must comply with the following conditions: (1) Live intensely a life of divine grace. (2) Prove your faith with words and actions, living a true Christian life. (3) Desire to remain under the protection of St. Padre Pio and to want to enjoy the fruits of his prayers and sufferings. (4) Imitate Padre Pio's virtue, particularly his love for Jesus Crucified, for the Most Blessed Sacrament, for the Madonna, for the Pope, and for the entire Church. (5) Be animated by a sincere spirit of charity towards all.

The prayer below is recommended for those who want to immediately become a spiritual child of Padre Pio with a written application to be sent to the monastery when feasible:

> O Saint Padre Pio, holy bearer of the Wounds of Christ, accept us this day as your spiritual sons and daughters and keep us always on the narrow path by your intercession. And do thou, O our Spiritual Father, stay there at the Gates of Heaven until all of your spiritual children have entered through, even and including us. Through Christ our Lord, Amen.

> Dear Padre Pio, I recall your promise to the Lord, "Lord, I will stand at the gates of heaven until I see all my spiritual children have entered." Encouraged by your gracious promise, I ask you to accept me as a spiritual child and to intercede for my prayer requests (Here state your petitions) Glory be to the Father, and to the Son, and to the Holy Spirit, now and forever.

Notes

[1]*Gente*, June 27, 2002, 21–25. Excerpted from http://www.sanpadrepio.com/promise.htm.

[2]https://caccioppoli.com/Padre%20Pio%20and%20his%20spiritual%20children.%20His%20words%20on%20meditation%20and%20prayer.%20Gift%20of%20tears.%20The%20Prayer%20Groups..html

[3]https://caccioppoli.com/Padre%20Pio%20in%20his%20own%20words%20about%20hymself,%20God,%20Jesus,%20Holy%20Spirit,%20Church.html

Moral of the Stories

"The virtues are what sanctify the soul, not supernatural phenomena." – *Padre Pio*[1]

At the center of Padre Pio's life was his love for God. His life was a life of total surrender to God dedicated to the mission of bringing all peoples to repentance and salvation.

This meant both constant prayer and meditation and endless hours in the confessional drawing people away from perdition and directing them to Paradise.

Padre Pio's love of God was manifested most powerfully in his love for the crucified Jesus and his burning desire to take up the cross and follow his Lord. The prayer he recommended for every need was the novena to the Sacred Heart of Jesus.

His love for Jesus was epitomized in his celebration of the Eucharist which was visibly a participation in the Passion.

As with all the saints, his love for God incarnate was accompanied by his great devotion to the Mother of the Savior for whom he had the greatest devotion. She was Jesus' gift to him – and to us!

Padre Pio recommended to all a spiritual regimen made up of weekly confession, daily communion, daily examination of conscience, daily spiritual reading and twice a day personal prayer.

Padre Pio's Five Point Regimen[2]

I. Weekly Confession

"Confession is the soul's bath. You must go at least once a week. I do not want souls to stay away from confession more than a week. Even a clean and unoccupied room gathers dust; return after a week and you will see that it needs dusting again!"

II. Daily Communion

"It is quite true, we are not worthy of such a gift. However, to approach the Blessed Sacrament in a state of mortal sin is one thing, and to be unworthy is quite another. All of us are unworthy, but it is He who invites us. It is He who desires it. Let us humble ourselves and receive Him with a heart contrite and full of love."

III. Examination of Conscience Every Evening

Someone once told Padre Pio that he thought a nightly examination of conscience was pointless because he knew what was sin as it was committed. To this, Padre Pio replied, "That is true enough. But every experienced merchant in this world not only keeps track throughout the day of whether he has lost or gained on each sale. In the evening, he does the bookkeeping for the day to determine what he should do on the morrow. It follows that it is indispensable to make a rigorous examination of conscience, brief but lucid, every night."

IV. Daily Spiritual Reading

"The harm that comes to souls from the lack of reading holy books makes me shudder.… What power spiritual reading has to lead to a change of course, and to make even worldly people enter into the way of perfection."

V. Mental Prayer Twice Daily

"If you do not succeed in meditating well, do not give up doing your duty. If the distractions are numerous, do not be discouraged;

do the meditation of patience, and you will still profit. Decide upon the length of your meditation, and do not leave your place before finishing, even if you have to be crucified. Why do you worry so much because you do not know how to meditate as you would like? Meditation is a means to attaining God, but it is not a goal in itself. Meditation aims at the love of God and neighbor. Love God with all your soul without reserve, and love your neighbor as yourself, and you will have accomplished half of your meditation."

God

Padre Pio's whole existence was centered on his relationship with God and he had much to say about God.[3]

"God loves man with an infinite love"
"God tolerates us, even when we offend him."
"God enriches a humble heart with his gifts."

"God goes in search of obstinate souls."
"God is terrible justice and infinite mercy."
"God will always give us more than we deserve."

"Time spent for the glory of God is never wasted."
"Life without the love of God is worse than death."
"Always remember that God sees everything."

"My past, O Lord, to your mercy; my present, to your love; my future, to your providence!"
"If God would take back what he gave us, we would be left with our rags."
"God is always fixed in my mind and imprinted in my heart."

"As long as one has God as a foundation everything else falls into place."
"When God punishes He does so with reverence, almost fearing to hurt."

"Confidence and fear of God go together like two sister."

"Let's look beyond the hand of men, to find the hand of God."
"Honor and praise the Creator and not the creature."
"Endure tribulation, sickness and pain for the love of God."

'Don't wait for Tabor to see God: you already see Him on the Sinai and on the Calvary."

Jesus

Padre Pio's life was a participation in the Passion of Jesus. All his actions were driven by the mission of Jesus.

We might ask who is Jesus of Nazareth and why is he relevant to our lives? Jesus himself asked: "Who do you say that I am?" (*Matthew* 16:15).

Jesus' followers through the centuries have claimed about him that he is God and man, that: Jesus is the human locus of the divine, the human face of God; the incarnation of the "Second Person" of the Triune God; the Eternal Word made flesh. He is the incarnation of Infinity.

Oxford thought-leader, Richard Swinburne, said in his book *Was Jesus God?* (published by Oxford University Press in 2008), "We would expect God (if there is a God) to become a human prophet and lead the kind of life and give the kind of teaching considered [in the book] … and to put his signature on that life by a miraculous event, such as a resurrection of that human prophet from the dead. … It is not merely the case that Jesus is the only serious candidate in human history about whom we have evidence that he lived the right kind of life which ended with a divine signature. Jesus was both the only prophet in human history about whose life there is good historical evidence of the first kind (evidence that he or she lived a perfect life with much suffering, claimed to be divine, claimed to be making atonement, gave plausible moral and theological teaching, and founded a church to continue his work), and also the only prophet about whose life there is good historical evidence of the second kind (evidence that his or her life ended with a miracle recognizable as a divine signature)."[4]

And BECAUSE Jesus is God and man, he is

- the Redeemer of the human race who has liberated us from a primordial curse
- the Savior who delivers us from our enslavement to evil and the danger of final damnation
- the transmitter of the very Life of God.

Consequently, Jesus is nothing less than the fulcrum of human history and human destiny.

The claim that Jesus is God and man is not simply a claim about the identity of Jesus but of the Godhead Itself. Jesus' life and being – the Jesus Cloud – is a manifestation in human terms of the innermost reality of the Godhead as Trinity: the Three "ways of existing" (hypostases) of infinite-eternal Spirit. As is evident in the Gospels, Jesus' identity is inextricably "related" to the Father and the Holy Spirit. And by his teaching, Jesus confirmed the pre-Christian intuition that God is indeed tri-personal.

Jesus presented a new and unprecedented vision of reality:

- God is Spirit
- God is Trinity (Three "centers" united in One spiritual reality)
- God (specifically the Second Person of the Trinity) is incarnate in him: he is the human locus of the divine who manifests through his life the tri-personal Being of God
- He, Jesus, atones for the sins of humanity
- He is the path to eternal life
- He enables our souls to be sanctified
- He lays out an ecosystem of self-giving love

This was the Jesus adored, loved and lived by Padre Pio. Hence, he said,[5]

- "Without Jesus you can do nothing."
- "On Calvary Jesus redeemed us, and salvation must be accomplished there."
- "I live for Jesus Christ, I live for his glory, I live to serve him, I live to love him."
- "In Jesus everything has an answer. Without him - only a big void."

- "It is to your advantage to commit yourself to the sufferings that Jesus will send to you. Jesus cannot tolerate seeing you afflicted and he will come to you and comfort you, blessing you with many graces for your soul."
- "I will never tire of praying to Jesus."
- "Life is a Calvary, but we must climb it cheerfully."
- "From the Calvary we go to the Tabor."
- "Jesus, let me suffer, and let me die from suffering."
- "I suffer only when I don't suffer."
- "I want to suffer, but in secret."
- "I desire nothing else than to love and suffer"
- "I want to save souls by continuous prayer."

The Eucharist

The story of the Eucharist is the amazing announcement that the very Life of God is offered to us in a medium customized to our physicality and oriented to our transcendent destiny. The Eucharist makes sense of both the religious rituals of humanity and the very history of life in the Universe. It is the climax of both the history of religious worship and the history of the progressive elevation of matter. This is so because it is the transformational point in space and time when the physical becomes a vehicle of the Divine Life.

The Eucharist, which is inclusive of the Eucharistic celebration, the Blessed Sacrament and Holy Communion, was Padre Pio's life-blood and life-force:

> As a young man, before he entered the priesthood, he spent hours in the church, adoring Jesus in the Blessed Sacrament. After his ordination, he took a long time for the Consecration of the Mass, to the point where parishioners complained about all the time he spent, in ecstasy, before the bread and wine as they became the Body and Blood of Our Lord Jesus. He had to control himself, to break out of his ecstasy, and force himself back to the Mass, at the orders of his superiors.

He began his preparations for Mass hours before the Mass began. He would constantly ask his fellow Friars what time it was. He always thought it was time to begin Mass, even though it might have been two hours early. When it came to celebrating the Eucharist, he was like a young race-horse, chomping at the bit, waiting impatiently to get out of the starting gate.

We read an eyewitness account in the Voice of Padre Pio, about Padre Pio and the celebration of the Mass.

"But as he started vesting for Mass, his body began to bend forward. As he slowly approached the altar, his body stooped over more and more as if he was being crushed under a heavy cross beam on his shoulders; his gait dragged, his face took on the sorrow of his Lord and Savior. As the mystery of the Sacrifice of the Mass unfolded, Padre Pio reached the pinnacle of suffering at the moment of Elevation of the Host in Consecration.

To quote a witness, 'In his eyes I read the expression of a mother who assists at the agony of her son on the scaffold, who sees him expire and who, choked with suffering, silently receives the bloodless body in her arms, able only to give slight caresses.'

He cried during the Mass. It was not weeping so much as it was deep, involuntary sobbing. He cried from the depths of his soul. When he beat his breast during the Confiteor, it was as if he was accusing himself of all the sins committed by man."

Giant tears cascaded from his closed eyes onto his beard. He took everything that was happening during that time so seriously. "Padre Pio's Mass put him into the drama of Calvary. For him it was reliving daily the pain which had wounded him in soul and body that morning of 20 September 1918. The Mass was his daily restigmatization."[6]

Padre Pio said about the Eucharist:

"The Mass is the complete union between Jesus and me."

"The holy Eucharist is a great means through which to aspire to perfection. But we must receive it with the desire and intention of removing from the heart all that is displeasing to him with whom we wish to dwell."[7]

"The Eucharist gives ad idea of the union we will have in heaven."

"How could I live failing even for a single morning to receive Jesus in the Communion?"[8]

Confession

Padre Pio accomplished his divinely ordained mission most fully in the confessional.

The divine Life is conferred at baptism, unleashed by confirmation and constantly nourished by the Eucharist. But the life of God departs from those who commit serious sin. Such sin is called mortal precisely because it inflicts a mortal wound. "All wrongdoing is sin, but there is sin that is not deadly (1 *John* 5:17). At the same time, "There is such a thing as deadly sin" (1 *John* 5:16). The sacrament that restores the divine life is Penance/Reconciliation and is administered in the confessional. The extraordinary power to forgive sins was conferred on the Apostles by the Risen Christ: "He breathed on them and said to them, 'Receive the holy Spirit. Whose sins you forgive are forgiven them, and whose sins you retain are retained.'" (*John* 20:22-3) This is a sacrament exercised through the power of the Spirit. Through it the penitent receives a fresh outpouring of the Holy Spirit.

In the normal course of life, we hurtle headlong into Hell. Mindlessly, helplessly, ever faster.

From our mother's wombs, through the hormonal furies of adolescence, to the seething perversions of adulthood, we seem marked for damnation:

By forces within and without: The Devil, the "world," weakness of will, passions and habits, instincts and impulses, sickness of soul.

But in the damning darkness a light shines. A power pivots us from our plunge to perdition. It is an open heart, the Heart of God. It is the Sacred Heart that was offered up on Calvary, that awaits us in the eucharistic Tabernacle and that calls out to us in the confessional.

All this the Padre knew and taught.

> As a confessor, Padre Pio wanted people to understand the seriousness of sin. "We have a greater fear of mortal sin than of fire," Padre Pio once said. On another occasion he said, "Beware of sin as of a poisonous viper." When penitents put questions to him regarding moral issues, his answers left no doubt as to the difference between right and wrong and the proper course to follow. One man said, "Padre Pio's words were firm, candid and pure." A man once confessed to him that he had thoughts against chastity. "How many times have you had those thoughts?" Padre Pio asked. "Six or seven times," the man replied. "But seven is not the same as six because it means one more deadly sin," Padre Pio answered.
>
> Padre Pio had a great fear of offending God and was ready to go to any length to avoid doing so. He had no illusions about human nature. He said, "As long as there remains a drop of blood in our bodies, there will always be a struggle between right and wrong." Looking back on his life, he once said, "Temptations that concern my secular life are those that most upset me ... They bring on a cold sweat and make me tremble ... In those moments, all I can do is cry."
>
> In 1915, Padre Pio wrote to Father Agostino: "The thought of going astray and ... offending God fills me with terror. It paralyzes my limbs, and both body and soul feel as if they are being squeezed in a powerful vise. My bones feel as if they are being dislocated .. . crushed and ground up."
>
> The general opinion was that making one's confession to Padre Pio was of profound spiritual benefit. Nevertheless, to confess to Padre Pio was not an easy task for most. As one person described

it, "To go to confession to Padre Pio was to allow him to look right inside your soul." As a confessor, he was strict and demanding. He had great moral strength in directing souls and he did not hesitate to tell the penitents what they needed to do in order to change their lives. He often told people what they did not want to hear. He had a strong character and was afraid of no one. Many people wanted to make their confession to him but were held back by their fear. One man stated, "It is less frightening to take a difficult examination at the university than to make one's confession to Padre Pio."

In the confessional, Padre Pio did not want people to make excuses for their sins and omissions. A woman from Gioia del Colle, Italy visited Padre Pio on one occasion. During her confession, she said that she missed Mass the previous Sunday because of the rain. "Yes, but when you left to come to San Giovanni Rotondo, it was raining too," Padre Pio replied. "You must never miss Mass again on Sunday unless illness prevents you from attending," he added.

An atheist was once introduced to Padre Pio and the visit resulted in the man's conversion. He said, "I went to see Padre Pio when I had a thousand reasons for not believing in God. With great delicacy, little by little, he led me back to the faith and gave me the moral stability I lacked."

Padre Pio attached enormous importance to the frequent reception of the sacrament of confession. He used to say, "Even if a room is sealed off completely, dust will still accumulate in it." Padre Pio practiced what he advocated to others. He went to confession frequently. Before making his confession, he prayed deeply and sought the intercession of the Virgin Mary. He always felt a great remorse for his sins and often cried when making his confession. To Father Benedetto, who was his spiritual director for twelve years, Padre Pio wrote, "I am seeking the amendment of my life, my spiritual resurrection, true and substantial love, the sincere conversion of my whole self to Him."[9]

As the great doctor of the soul, Padre Pio advises us[10]:

> "Before going to sleep examine your conscience and turn your thoughts to God."

> "The Lord allows us to discover who we are a little at a time."

> "The sinner who is ashamed of his bad deeds is closer to God than the just ashamed of acting well."

> "One day the slightest transgression of God's law will be considered."

> "Implore the Lord to increase two things in us: love and fear."

> "As long as you are afraid to fall, you will not sin. You should be afraid when you are not afraid anymore."

The Blessed Virgin Mary

The odyssey of the Blessed Virgin Mary is one of the most astonishing tales in the history of humanity. But it is so familiar we take it for granted. And so we fail to realize its utterly remarkable nature. Here was a human person whose Son was God incarnate – and whose mother she remains for all eternity. A person who entered into the most intimate union possible union for a human person with God the Holy Spirit. A person specially chosen and prepared by God the Father to be not just the Mother of his only begotten Son (*Luke* 1:30) but of all who bear witness to the Son and keep the divine commandments (*Revelation* 12:17). Most important, the Incarnation of God the Son was contingent on her consent – a consent she granted from the depths of her being. She was called to share totally in the sorrow that was to be his life (*Luke* 2:30) – a participation she whole-heartedly offered up. The earliest apprehension of her by the faithful was, in fact, in her role as Sorrowful Mother.

Our salvation comes from Christ Jesus. But the salvific coming of Christ was made possible by his Mother's free act of obedience to the divine Will. She was not a puppet, not a mindless vessel chosen by a super-human puppet-master to achieve his aims as some critics of Marian devotion have said. Such critics are ideologists who deny the plain words of Scripture and devalue both the dignity of the human person and the gift of freewill given by our Maker. But the followers of Christ from the beginning recognized that she was not simply chosen by God but that Mary herself chose God in the most important choice ever made by a human person. Therein lies her blessedness through all generations. "Blessed are *you who believed* that what was spoken to you by the Lord would be fulfilled." (*Luke* 1:45).

Like all the Fathers, Doctors, Mystics, Councils and Liturgies of the Church, Padre Pio was aware of the Virgin Mary's pivotal role in salvation history. She was the New Eve with Jesus the New Adam. We see that she is mysteriously present with the New Adam at precisely the times most crucial to the accomplishment of Redemption. When she offers up her infant Son at the Temple, it is prophesied that a sword will pierce her soul. This prophecy was fulfilled, said the Fathers and the faithful, when her offering came to a climax on Calvary and she became for all time the Sorrowful Mother. The consent given at the Annunciation extended through the Presentation at the Temple to the Sacrifice on Calvary.

As the old Eve was associated with the old Adam at all stages of the Fall, so also the New Eve was associated with the New Adam at every step of the road to Redemption. The Fathers recognized that the Incarnation cannot be separated from the Cross and Redemption and in calling Mary the New Eve they drew our attention to the singular role she played in the Redemptive Mission of God Incarnate.

She is permanently the New Eve just as her Son will always be the New Adam.

Padre Pio was "connected" to the New Eve from his earliest years:

> Padre Pio was 14 years old when he visited the shrine in Pompeii dedicated to Our Lady. Our Lady of Pompeii was especially dear to his heart and often when he needed a special grace for himself or for someone else, he would ask his friends to pray to Our Lady of Pompeii for his intentions. He wrote, "I should like to ask you, if it is not inconvenient, to do me the kindness of making three

consecutive novenas to the Virgin of Pompeii for a grace to be obtained for me from her Son, a grace which will mean a great deal for a certain soul." On another occasion he wrote, "I thank you for the novenas you made for me to Our Lady of Pompeii and I would ask you if it is not inconvenient, to continue because I am in great need."

In his letters to his spiritual directors, Padre Pio called the Blessed Mother, "beautiful Virgin Mary," "most tender mother of priests," "Mediatrix of all graces." In Our Lady, he saw the advocate of sinners, the most beloved, the consoler. He frequently referred to Mary simply as "Mother." Reciting the Angelus he rarely managed to restrain his emotions and when he spoke about her it was not unusual for him to shed tears.

The Sorrowful Virgin was another attribute of Mary that was close to his heart. He loved to contemplate her at the foot of the cross. To his spiritual children he would say, "Go and keep company with Jesus in His Passion, and with His Sorrowful Mother."

From the day on which he became a priest he always preferred to say Mass at Our Lady's altar. The Mass he celebrated most frequently was the Mass of the Immaculate Conception. "The Immaculate Conception," he said, "is the first step on the path of salvation."

To a spiritual daughter who asked him for a thought on the Madonna, he said, "My daughter, it is enough for you to know that Mary is the Mother of Jesus. . .She loves us so much that she offered to God the Father His only natural Son to save His adopted sons. . .She is a great and inestimable treasure who encloses in herself an infinite treasure, the Son of God."

Padre Pio used to speak of being inundated with graces by her. "She treats me as if I were her only child on the face of the earth," he wrote. He spoke of the tenderness of Our Lady who accompanied him to the altar. "What care she took to accompany me to the altar this morning. It seemed to me that she has nothing else to think about except myself as she filled my whole heart with sentiments of holy love."

Padre Eusebio Notte, personal assistant to Padre Pio, once said to him, "You see Our Lady, don't you?" He answered, "She comes to me whenever I need her." He also said that she remained beside him when he heard confessions. "Her loving care toward me cannot be described in words," he said.

Padre Pio's love for Mary was evident as he prayed the Rosary whenever there was a free moment in his busy day and when there were no free moments, he found a way to pray even then. He preferred the Rosary above all other prayers and asked his spiritual children to carry a Rosary with them at all times and to pray it every day. For him the Rosary represented the continual meditation on the profound mysteries of God's designs for salvation.

Padre Pio said so many rosaries in one day that the number was unbelievable, although he had a special grace for doing this. His spirit of prayer made what seemed impossible, within his reach. "The Lord asks this of me; He does not ask the same of you," he said. Once someone asked him how he was able to pray so many rosaries. "You can do one thing at a time," he replied, but I can do three or four."

Padre Pio said to one of the friars, "My son, if we do what we have always done, what our fathers did before us, we cannot go wrong. Satan wants to destroy this prayer, but in this he will never succeed. The Rosary is the prayer of those who triumph over everything and everyone. It was Our Lady who taught us this prayer, just as it was Jesus who taught us the Our Father." Padre Pio defined the Rosary as "The synthesis of our faith, the expression of our charity, and the foundation of our hope."

Daniel Hickey, from St. Albans, New York, was an American G.I. stationed in Italy during World War II when he met Padre Pio. When he returned to the U.S., Danny entered the seminary and became a Capuchin priest. Recalling his visits with Padre Pio he wrote, "Padre Pio was usually seen standing with his right hand in the front fold pocket of his Capuchin habit. A few times when he withdrew his hand, he was seen to be fingering a small chaplet of beads. It seems it was Padre Pio's habit not to waste a second but to fill each one with a prayer. Any lull in the conversation, no matter how short, his lips would be seen to move slightly as he prayed. But there was nothing ostentatious in all this; it was a long time before I was aware of what he was doing, though I had noticed his hand constantly in the breast pocket of his habit."

Father Alessio Parente said, "I was at his side for six years, and in all that time I never saw him without the Rosary in his hands night and day. Our Lady never refused him anything through the Rosary. The Rosary was his constant link with Our Lady.

Not only did Padre Pio have a profound devotion to the Mother of God, but he also seemed to know the dispositions of her heart. An image of the Madonna was found to be weeping in one of the parish churches in Italy. The bishop of the diocese asked for an investigation into the matter, and after much study the bishop declared it to be authentic. When Padre Pio was told about this he said, "When Our Lady is weeping, things are not going well."

Bill Martin (who later became Father Joseph Pius Martin) from Brooklyn, New York, first met Padre Pio in 1959. It was during his second visit in 1964 that he found it very hard to leave. As he was waiting for the bus to take him to Foggia, to his great relief a friar came running to tell him that Padre Pio had sent him to say that he wanted Bill to stay. He entered the friary as a 3rd order Franciscan and was known to everyone as Brother Bill. He became Padre Pio's personal assistant. One afternoon as he was sitting next to Padre Pio on the balcony near Padre Pio's cell, he was thinking to himself how very fortunate he was that he had been able to be so close to Padre Pio through the years. He wondered to himself who had obtained this marvelous grace for him. No words passed between the two but at the moment that Brother Bill was thinking those thoughts Padre Pio turned to him and said, "It was Our Lady."

The Virgin Mary, ever close to Padre Pio, assisted him in a special way regarding the state of his health. Although Padre Pio suffered from poor health his entire life he did not become discouraged. He offered all of his sufferings to the Lord for the conversion of sinners and for the souls in purgatory. He said that God had "put the problem of my health and a victorious outcome into the hands of our heavenly Mother." Once when he was ill, he confided to his Superior, "What made me suffer more than anything was not being able to say even one Ave Maria."

Not only was the Virgin present to Padre Pio in times of physical infirmity but in times of interior trials and spiritual suffering as well. On August 15, 1929 on the Feast of the Assumption, Padre Pio described an experience of Our Lady while he was celebrating Mass. "This morning I went up to the holy altar I know not how. Physical pain and interior grief competed as to which could most afflict all my poor being… A mortal sadness pervaded me through and through and I thought that all was finished for me…. At the moment of consuming the Sacred Species of the Host, a sudden light flooded through me and I clearly saw the Heavenly Mother with the Christ Child in her arms who together said to me, "Stop worrying! We are with you, you belong to Us and We are yours."

Pio testifies to the Virgin Mary's help in a number of letters to his spiritual directors. On May 26, 1910 he wrote to Father Benedetto, "My only regret, dear Father, is that I have no adequate means with which to thank the Blessed Virgin Mary, through whose intercession I have undoubtedly received so much strength from the Lord, to bear with sincere resignation the many humiliations to which I am subjected day after day…and I do not believe this strength comes to me from the world."

There were always crowds of people around Padre Pio, especially the sick, trying to get close to him and to touch him. To the sick and the suffering who implored his help, the Padre would say, "Entrust yourself to Our Lady." He would often say, "Let us pray to Our Lady that she snatch this grace for you from the Heart of Jesus." And for those who came to him with especially difficult personal problems, family problems, etc. he would often say, "Here we need the Madonna."

Cleonice Morcaldi, one of Padre Pio's spiritual daughters, spoke to Padre Pio on September 20th, just three days before he died. "Father, give me at least one word," she said to him. "Love the Madonna and make her loved. Always recite her Rosary. That is an armor against the evils of the world today." Cleonice asked him, "Is the Madonna close to you?" "A Mother..." he replied. "All of paradise is near her."

Padre Pio died on September 23, 1968. Padre Pelegrino Funicelli, his brother in religion, assisted him in his last moments on earth. He said that a short time before Padre Pio's death and with great tenderness, Padre Pio gazed at a picture of Our Lady Liberatrix that was hanging in his cell. He passed into eternal life peacefully, gently. He died with his Rosary. His last words were Gésu, Maria, – Jesus, Mary which he repeated over and over until he breathed his last.[11]

Notes

[1] *The Voice of Padre Pio*, Vol. XXII, No. 4, 1992, 12.

[2] https://joyfilledfamily.com/2021/08/spiritual-duties.html

[3] https://caccioppoli.com/Padre%20Pio%20in%20his%20own%20words%20about%20hymself,%20God,%20Jesus,%20Holy%20Spirit,%20Church.html

[4] Richard Swinburne, *Was Jesus God?* (Oxford: Oxford University Press, 2008), 128.

[5] https://caccioppoli.com/Padre%20Pio%20in%20his%20own%20words%20about%20hymself,%20God,%20Jesus,%20Holy%20Spirit,%20Church.html

[6] https://www.truechristianity.info/en/articles/article0013.php

[7] https://www.catholicdigest.com/from-the-magazine/quiet-moment/st-pio-of-pietrelcina-the-holy-eucharist-is-a-great-means-through-which-to-aspire-to-perfection/

[8] https://devoutlycatholic.wordpress.com/2013/04/19/daily-reflections-of-saint-padre-pio-on-the-eucharist/

[9]https://padrepiodevotions.org/pray-hope-dont-worry-issue-62-january-march-2015/

[10]https://caccioppoli.com/Padre%20Pio%20in%20his%20own%20words%20about%20hymself,%20God,%20Jesus,%20Holy%20Spirit,%20Church.html

[11]https://padrepiodevotions.org/pray-hope-and-dont-worry-issue-21-july-september-2004/

APPENDIX

Sadducees and Pharisees

Some theories are so "preposterously silly that only very learned men could have thought of them. But such theories are frequently countenanced by the naive since they are put forward in highly technical terms by learned persons who are themselves too confused to know exactly what they mean."[1]
Philosopher C.D. Broad

"I really do believe that our attitudes are shaped much more by our social groups than they are by facts on the ground. We are not great reasoners. Most people don't like to think at all, or like to think as little as possible. And by most, I mean roughly 70 percent of the population. Even the rest seem to devote a lot of their resources to justifying beliefs that they want to hold, as opposed to forming credible beliefs based only on fact. Think about if you were to utter a fact that contradicted the opinions of the majority of those in your social group. You pay a price for that. ... The decisions we make, the attitudes we form, the judgments we make, depend very much on what other people are thinking. ... One danger is that if I think I understand because the people around me think they understand, and the people around me all think they understand because the people around them all think they understand, then it turns out we can all have this strong sense of understanding even though no one really has any idea what they're talking about."[2]
Cognitive scientist Steven Sloman

> *"One has to belong to the intelligentsia to believe things like that. No ordinary man could be such a fool."*[3]
> George Orwell

Despite all the data testifying to the authenticity of the Padre Pio phenomenon, there will still be those who remain skeptical. Padre Pio faced skeptics from the beginning. Some were the village atheists. Others were his ecclesiastical antagonists. The first were the Sadducees and the second the Pharisees.

The critiques of the skeptics often told you more about the skeptic's mind-set than the content of their charges. Those who dogmatically attribute all apparently miraculous phenomena to hysteria often appear to suffer from an inverse version of hysterical personality disorder. Those who regard all such phenomena as products of clerical conspiracies tend to be prone to conspiracy theory-thinking. They see conspiracies where everyone else sees events proceeding on their own steam.

Plausibility matters little to the skeptics' "and then no miracle happened" school of thought. The only objective is to stop the barbarians at the gate – the hordes of sensationalist supernaturalists and pious Puritans. We should use all the means at hand to keep the knights of faith at bay. Coincidences galore, natural agents that perform supernatural acts, spectacular feats of speculation, whatever it takes.

But there is no reason why the rest of us need to imprison ourselves in the skeptics' mental cages. No matter how entertaining all this may be, at some point we have to leave the circus tent. There's a real world out there where the tigers are not jumping through hoops and acrobats are not leaping from one trapeze to the next. This real world operates with rules of evidence and explanation. It assumes above all the reign of sanity. Flights of fancy, silly seasons, freak shows – these are barnacles hanging onto a ship not the ship itself. So it is with the inquiry into the Padre Pio phenomenon. We should follow the evidence where it leads.

Stigmatizing Stigmata

With regard to the stigmata, we have seen why the two favorite arguments of the skeptics fail – the carbolic acid and the psychosomatic explanations.

Here we will touch on a few other more juvenile charges.

Joe Nickell, the critic-at-large of the paranormal realm, as is his standard practice, simply assembles all the usual discredited critiques of Padre Pio neither offering any of his own nor acknowledging the well-known refutations of these critiques. He does, however, add this curious comment: "Eventually Pio began wearing fingerless gloves, supposedly to cover his stigmata out of pious humility; however, to me, the practice seems instead a shrewd move to eliminate the need to continually self-inflict wounds."[4] He seems entirely unaware that, first, Padre Pio wore those gloves on the instructions of his superiors and, second, his bleeding hands were always exposed during the celebration of the Mass.

As for refuting the reported miracles of Padre Pio, Nickell simply cites the (again discredited) charges of the clerical enemies of the Padre – from 1919! In this case the Sadducee needed the Pharisees!

Some skeptics point to the fact that Padre Pio bled from his palms whereas current research indicates that nails in a Roman crucifixion were driven into the wrists. Here the analysis from Dr. Frederick Zugibe, a forensic pathologist and cardiologist who was Chief Medical Officer of Rockwood County, New York (1969 to 2003), Adjunct Associate Professor at the Columbia University College of Physicians and Surgeons and Director of Cardiovascular Research with the US Veterans Administration, will be of help. Dr. Zugibe was considered one of the leading forensic pathologists in the US and directed some 10,000 autopsies. In his own modeling of the crucifixion, Dr. Zugibe shows that it is not an either/or question with regard to wrists or palms because the nails could enter through one area and exit through another. For instance, according to Zugibe, "the Shroud of Turin only shows the site of the nail's exit and not where the nail entered. There are only two possibilities as to where the nail entered: through the radial side of the wrist or through the upper part of the palm angled toward the wrist. The most plausible region for the nail entry site in the case of Jesus is the upper part of the palm since this area can easily support the weight of the body, assures no bones are broken, marks the location where most people believed it to be."[5]

There is no absolute requirement that the wounds of a stigmatist have to be in the exact location of the wounds of Jesus. But, as Zugibe shows, the palms vs. wrists dilemma has no bearing on the Padre Pio stigmata.

It is also said that in his initial description to his superior of his bewildering experience, the young Padre Pio used the words of another stigmatist. It is hard to tell what is the problem here. If someone who suffers a bullet wound accurately describes the pain they feel at that moment, there is no reason why someone else with the same wound cannot use that existing description in writing about the felt experience to a friend. Why re-invent the wheel? In trying to understand what was happening to him, the young Pio, with his spiritual director's permission, turned to the writings of a contemporary, Gemma Galgani, who had undergone similar trials. Also, his correspondence on this matter was supposed to be private and Padre Pio was mortified when it was later published. In retrospect, it is evident that his experience was part of the "dark night of the soul" described by such great mystics as St. John of the Cross. Of course, descriptions are one thing. Actually experiencing what is being described is a different matter. After the initial shock, Padre Pio lived with the wounds in all their agony for 50 years.

Some historians and sociologists have suggested that socio-economic and geopolitical events could trigger stigmata. As it pertains to Padre Pio, they say, Italy had just engaged in a traumatic war and this might have brought about his reporting of the stigmata. About this type of reductionist "explanation," all we can do is cite Orwell's comment: "One has to belong to the intelligentsia to believe things like that."

We should note here that Padre Pio is not responsible for the actions of the inhabitants of San Giovanni Rotondo some of whom, in the first days of the stigmata, apparently sold pieces of clothing with chicken blood as "relics." Padre Pio himself refused to give anyone his bloodied clothing except when it was needed for studies sanctioned by his superiors.

Guilty Until Proven Guilty

Where the Sadducees sought to expose the stigmata and miracles as natural phenomena mixed with fraud, the Pharisees sought to go further through a campaign of character assassination and demonstrably false slander. Both groups used each others' fabrications to further their own agenda.

Padre Pio was always in the public eye. There were three principal places he could be found: in the confessional, celebrating Mass in the monastery church or praying in his cell. Wherever he walked he was surrounded by the faithful or by his fellow monks. The only place where he could be alone was in his cell, a place where no women were allowed entry. So the salacious charges hurled at him by his enemies always sounded preposterous on the face of it.

The origin of these scurrilous charges may be traced to his ecclesiastical antagonists who trafficked in gossip and little else.

For a third party history of the provenance and nature of these charges, we turn to the comprehensive biography of Padre Pio authored by a Lutheran, C. Bernard Ruffin. Ruffin made it clear that his was not a work of hagiography.

One of Padre Pio's earliest and fiercest critics was Archbishop Pasquale Gagliardi who was, from 1897 to 1929, the Archbishop of Manfredonia, the archdiocese in which San Giovanni Rotondo was located. Gagliardi had powerful friends in the Vatican and Pope Piux XI, in fact, relied on him for his assessment of Padre Pio. Archbishop Gagliardi disliked the Franciscans in general and Padre Pio in particular.

Ruffin writes,

> Gagliardi encouraged his priests to bombard the Holy Office with lurid accusations concerning the ministry and character of the stigmatized priest. (...) Many years later, the Capuchin Padre Alberto d'Apolito interviewed Don Domenico Palladino, a priest from San Giovanni who had been one of Padre Pio's "most venomous detractors" [asking him why he lied and] denigrated

and calumniated Padre Pio (…) [Palladino replied] "At that time I had to obey the one who gave me orders." When asked who gave him these orders, Palladino replied, "The bishop. I was very young to reflect and think about what I was doing. I knew I had to obey."[6] (…)

The members of a particular convent of Poor Clare nuns complained that Gagliardi on occasion entered the cloister to spend the night with a niece of the abbess. A priest insisted that he had surprised the archbishop in bed, naked, with another nun. Moreover, it was widely known that some of the clergy, like Prencipe and Palladino, openly lived with their mistresses with the archbishop's knowledge. Worse, Gagliardi protected priests who were child molesters. [7] (…)

Padre Pio aroused the fury of San Giovanni's archpriest, Don Giuseppe Prencipe, by denying absolution to his mistress, Maria DiMaggio. Concubinage, like homosexuality, seems to have been rampant among the clergy in the Manfredonia archdiocese, and Don Giuseppe and DiMaggio had lived together for more than twenty years. When DiMaggio came to him for confession, Padre Pio declared that her relationship with the archpriest was sinful and had to be terminated at once if she wanted absolution from her sins. (…) Through it all, Archbishop Gagliardi supported Prencipe and once more denounced Padre Pio for his "horrible means of hearing confessions."[8]

When one of his key cardinal-protectors in the Vatican died, Gagliardi was forced to resign in 1929. Although his successor as Archbishop of Manfredonia became one of Padre Pio's great defenders, others stepped in to continue the relentless campaign of slander and persecution.

One such vicious critic of Padre Pio was Bishop Carlo Maccari who was sent to investigate Padre Pio by Pope John XXIII in 1960. Ruffin writes that Maccari was described as "hard, authoritarian and unyielding". He brought with him as assistant a priest called Giovanni Barberini.

"Maccari summoned all the surviving members of the faction that had attempted to undermine Padre Pio more than three decades before. De Nittis was overheard remarking to Don Domenico Palladino, who had been one of Padre Pios most unyielding detractors, "Well, Dumi, the hour of revenge has come!""

Padre Raffaele D'Addario, a classmate and close friend of Padre Pio, observed about Maccari, "From the way he talked he seemed to want to let me understand that he thought that everything connected with Padre Pio was a show and foolery."

Maccari's assistant Barberini was "visiting restaurants, bars, and shops, trying, far into the night, to learn what people thought of Padre Pio. Padre Raffaele related that Padre Pio was repelled by Barberini (who later left the priesthood) and tried to avoid him whenever possible. ... "Puppets of Satan," [Cleonice Morcaldi, one of the women interviewed by the two] called Maccari and his secretary." She said that Barberini "went to Casa Sollievo to fool around with the young women."

Relations between Padre Pio and Maccari got off to a decidedly rocky start. (...) Clearly neither man liked each other. ... Padre Pio remarked to a friend (...) '[He] wants to crucify me. Nevertheless I want him with me in paradise.' Although Padre Pio desired his salvation, he didn't like or trust Macari. And the distrust was mutual.

Maccari criticized Padre Pio for building his hospital in San Giovanni instead of Rome. "A month after you're dead, your hospital will have to close its doors."

Maccari wrote that he "could not deny that for years Padre Pio had been "an instrument of mercy, pardon and peace." He could not understand why God would give such a mission to such a "small and petty person." With the disdain that many northern Italians had for those from the south, he described Padre Pio as poorly educated and theologically unsound, due to his "southern origins." (...)

Maccari wrote a 208 report on Padre Pio for Pope John XXIII that was extremely critical. But the cardinals who knew Padre Pio defended him with the Pope.

> Cardinal Guiseppe Siri (1906-1989), Archbishop of Genoa said, "For months I defended Padre Pio to Pope John XXIII ... The pope was ... worried and concerned about what was reported to him." Siri pointed out to the pontiff that Padre Pio was the victim of slander. Pope John turned to his old friend, Archbishop Cesarano of Manfredonia, and asked him about Padre Pio when the prelate came to Rome. He told the pope, "Padre Pio is always the man of God whom I have known since the beginning of my [tenure as archbishop in] Manfredonia. He is an apostle who does immense good for souls." When the pope brought up the evil reports, Cesarano objected, "It's all slander. I've known Padre Pio since 1933 and I assure you that he was always a man of God."

> "Then those women, those recordings," the pope went on. "They even recorded kisses."

> "For charity's sake," Cesarano insisted, "it's not the case of sinful kisses." He explained how women continually grabbed Padre Pio's hand and kissed it, despite his complaints that this hurt his stigmata. Cesarano even observed his own sister kissing Padre Pio's hand over and over again.

> Cesarano convinced the pope that [the photographs of Padre Pio with women] had been doctored.

> About Elvira Serritelli, "the archbishop told the pope that it was well known that she was mentally unstable and not a trustworthy witness. As a result of this "and the defense of Padre Pio from other churchmen that he respected", the pope "did not follow through on the devastating recommendation to ban Padre Pio from saying Mass."

> Some years later, in a letter to the Holy Office, Maccari wrote, "I realize – at a distance of years and with an experience I hope more mature, of men and things – that my attitude towards Padre Pio

may have played a part in the way I proceeded," and he conceded that he might have been "too frank and harsh." *By the time Maccari died in 1997 (...) the aged prelate, who had become archbishop of Ancona (...) was referring to Padre Pio as a "saint."*[9]

About Maccari's accusations, Father Gerardo di Flumeri writes:

> His [Padre Pio's] pious penitent women were all elderly, all virtuous and also all devoid of any physical attraction. Yet all were questioned by Monsignor Maccari with the brutal question: did you make love with him? How many times?[10]

Writes Ruffin, "One of the local women, horrified, complained to Padre Raffaele, "Only a devil from hell would ask the kind of questions he did!" When questioned about Padre Pio's chastity, Padre Raffaele cut Maccari short, asserting, "All I can say is that, if all priests, both secular and religious, prelates included, had the baptismal purity and innocence of Padre Pio, the Church would be truly holy in all her members! I have nothing more to say."[11]

Unquestionably, the women targeted by Maccari had made enemies through their overall behavior. Ruffin quotes one of the Capuchins on this matter: "There were many spiritual daughters who Padre Pio directed. The greater part put his teachings into practice with a more perfect life. One group created an uncomfortable and embarrassing situation with their overbearing and domineering behavior. There were three women in particular in this group: Cleonice Morcaldi, Countess Telfner and Elvira Serritelli. Each of them insisted that she was the best loved and therefore they became officious and insolent. (...) They even had their designated places in church and they were bullies because they wanted to command. They were deaf to the recommendations [of Padre Pio] to be calm and respectful. They were hostile to the friars and in particular to the Provincial."[12]

With friends like this, who needs enemies, we might say. Ruffin continues:

> While Cleonice Morcaldi and her close friends, Countess Telfner and Tina Belloni, who claimed for themselves the right to sit on the first row in church when Padre Pio was celebrating Mass,

could be pathologically possessive of Padre Pio, the Serritelli sisters, Elvira and Angela, could be furiously aggressive. "Bitter and venomous," "aggressive, overbearing, and dangerous," and "hysterical, deranged, mentally ill," were adjectives frequently used to describe these unmarried schoolteachers. (...) Cleonice was so afraid of Elvira and Angela that, when they swept into church, she would hide in a corner. Padre Alberto recalled a time when he sat beside Padre Pio, while he was praying in the choir, overlooking the altar area. "Elvira, down there on the steps of the altar, with a vase of flowers in her hands, raised her gaze towards Padre Pio" who, "from time to time in the fervor of prayer, opened his eyes to gaze at the tabernacle." She imagined he was gazing at the cleaning ladies, rather than at her. Imagining that Padre Pio preferred the cleaning ladies to her, Elvira "became unnerved, shaking and trembling with rage. At that moment she would have been capable of any insane act." Maria Pennini, another schoolteacher, described Elvira Serritelli as a "very terrible, vindictive woman" who "went into rages and couldn't bear for anyone else to speak to Padre Pio." (...)

As he aged, Padre Pio lacked the energy to repel the aggressions of the "Pious Women." The other friars were appalled as these women strong-armed their way into Padre Pio's presence, demanding various blessings and favors, and insisted on long audiences in the guest room, where they drove Padre Pio almost to tears with their mindless and petty chatter. (...) When asked why he allowed these strident groupies to monopolize so much of his time, Padre Pio replied to the effect that they would give him no peace until he gave in. However, Padre Pio also believed that they were among those whom God had entrusted him to direct. (...)

Padre Pio's brother Michele [said] "Do you know what my brother's greatest cross is? Do you think it's the stigmata? It's not. It's those dreadful women."[13]

This new series of attacks was partially motivated by the success of the hospital begun by Padre Pio:

> In the sixties the very success of Casa Sollievo gave Padre Pio major difficulties. The steady pouring of contributions became a source of contention in the Capuchin Order.
>
> The Capuchins had fallen prey of the pyramid scheme financier Giambattista Giuffre'. The so called "God's Banker" gave interests from 30% to as high as 90%. The Capuchins, like many others, invested large sums of money with Giuffre'. Padre Pio, who had the ultimate administrative control over the funds, was approached by Capuchin functionaries, and asked to invest. Padre Pio refused. In the summer of 1958 Giuffre' declared bankruptcy. The Capuchins faced an economic disaster.
>
> They approached Padre Pio once again, and he once again refused to give the money donated to "Casa Sollievo'. At that point a plan emerged to take control of the hospital from Padre Pio, accusing him of being a poor administrator, disobedient, and morally unfit. [1] [Rega, F. M. (2005). Padre Pio and America. Rockford: TAN books, 235.]
>
> The plan included bugging the convent with tape recorders and microphones, in order to get confidential financial information especially from the administrator of Casa Sollievo Angelo Battisti, and to document inappropriate behavior between Padre Pio and his spiritual daughters. [2] [Giannuzzo, E. (2012). San Pio da Pietrelcina. Il travagliato persorso della sua vita terrena. Book sprint edizioni, 366.] (...)
>
> Padre Clemente: "Just like any other individual religious Padre Pio had a vow of poverty. Pope Pius XII did not dispense Padre Pio from the wow in general, But the Pope Pius XII said: "You can do everything with the total income of Casa Sollievo." Therefore Padre Pio was always under the vow of poverty, and at the same time he could dispose of the hospital funds and goods as he saw fit. The only suspension from the vow of poverty pertained to

the hospital Casa Sollievo."[6] [Schug, J. O. (1987). A Padre Pio Profile. Petersham, MA: St. Bedès Publications, 106.] (...)

On May 11, 1960 the first microphone was installed on the internal window separating the convent from the visitor's room. The conversation between Padre Pio and his spiritual daughter was recorded. Padre Giustino and fra Masseo, listening to the recording thought to hear "tender expressions" and "a kiss". The Holy Office was informed, and the tape brought to the Vatican, and the tape was reheard in the Holy Office, and Padre Giustino was summoned to the Holy Office "to better interpret the words pronounced in Neapolitan dialect".[7] [Giannuzzo, E., 371]

Padre Amedeo da San Giovanni Rotondo who heard the tape with don Terenzi reported: "The voice recorded was so faint that I could catch only few words, and never a complete sentence. And I paid attention but didn't hear any smacking of a kiss."[8] [Ruffin, C. B. (1991). *Padre Pio: the true story*. Huntington, Indiana: Our Sunday Visitor, Inc., 358.]

The other microphones were placed in Padre Pio's cell, and also in the confessional. Thirty six tapes were made during a three months period. [9] [Ruffin, 358.]

Padre Pio discovered himself a microphone under his bed, with the wire passing through the wall of his cell, to the cell of Padre Giustino. He cut the wire with a pocket knife.[10] [Giannuzzo, E.. 385-6.]

When Padre Pio discovered the wires he told his friend Francesco Morcaldi. Morcaldi leaked the story to the press.[11] [Ruffin, 358.]

Padre Pio said to Bishop Cesarano: "My own brethren are doing this to me."[12] [Ruffin, 358.][14]

It was these greed-motivated fabrications that led to the Maccari visitation.

About the tapes used by Maccari, Ruffin writes, "Most of the tapes were apparently inaudible; the ones that were audible recorded mundane and innocuous business."[15]

Padre Eusebio pointed out that "It was well known that kissing Padre Pio's hand was the order of the day and under the eyes of every one and on the part of everyone."[16]

In his study of Luzzatto's overview of the Maccari affair, Richard Demma writes,

> However, towards the very end of the book, the author devotes some pages to the most salacious accusations of all – the alleged evidence of secret microphones planted in various places, which 'seemed' to suggest some impropriety on Pio's part with his female followers. The content of these tapes apparently shocked Pope John XXIII (who didn't actually listen to them) and resulted in the final Vatican investigation, headed by Monsignor Carlo Maccari, who would later suggest that Pio may have been enjoying carnal relations with some of his female devotees as much as twice a week. Now the Maccari affair (grotesque and repellent, in my opinion- letting my own biases show) is something I know a little about, having researched it some years ago. And here is where I can certainly fault Luzzatto and my suspicious began to tilt in Tornelli's direction. **Luzzatto does not clarify that the microphones were not planted in Padre Pio's bedroom or the women's confessional, but only in the men's confessional and various visitors rooms where Padre Pio would converse with pilgrims. So in other words, we are not dealing with tapes that actually record intimate private moments between Padre Pio and women but only public conversations and bits of gossip from visitors in the rooms awaiting his arrival. Now that is a vital omission in a historical work purporting to be objective. It is the one detail that changes everything.** Instead, Luzzatto drops the general insinuation of 'secret tapes' and leaves it hanging, dripping with innuendo. Furthermore, 'news' (from official Capuchin sources) stated that Msgr Maccari had recanted his accusations on his deathbed and asked for Padre Pio's forgiveness and blessing. And that would be a major story, in itself a deathbed recantation from Padre Pio's most recent official examiner. Is this an apocryphal story or can it be objectively verified? At the very

least, if Luzzatto was as 'objective' as he claimed to be, then this incident should have been reported and explored - at least as to its plausibility. But not a word from Luzzatto (...) Unfortunately, these key omissions cast doubts on the reliability of the rest of the book, much of which does seem to me to be of great value. This leads me to believe that Tornelli may be right as to Luzzatto's ultimate intentions. So while the book did little to affect my own estimation of the great sanctity of Padre Pio, it did help me to understand the complexities of the human context within which he lived and worked - and it left me with some serious doubts as to Luzzatto's ultimate fairness and objectivity as a scholar.[17]

The attacks on the character of this holy man bring to mind the slanderous attacks on Jesus of Nazareth both when he walked this earth and in later centuries. If the calumnies uttered against Padre Pio were malicious and injurious, the blasphemies against his Lord and Master were infinitely more destructive and despicable. "No slave is greater than his master nor any messenger greater than the one who sent him." (*John* 13:16)

The Last Word

The skeptics we will always have with us. They have no barriers to entry, no accountability for mindlessly propagating egregious errors. But for those who wish to remain faithful to the facts and to be led by the evidence as we have it, the explanation of the Padre Pio phenomenon is obvious and undeniable. There can be no reasonable doubt that Padre Pio was not just a remarkable man but that Providence could hardly have chosen a more appropriate saint for the age in which he lived.

Notes

[1]C.D. Broad, *The Mind and Its Place in Nature* (London: Kegan Paul, 1925), 623.
[2]http://bit.ly/2m09ew
[3]George Orwell, *Notes on Nationalism*, May, 1945. http://bit.ly/1QiVrwG
[4]*Skeptical Inquirer*, March/April 2011, 59.
[5]https://www.shroud.com/zugibe.htm

[6]C. Bernard Ruffin, Padre Pio: The True Story (Huntington, IN: Our Sunday Visitor, 2018), 207-8

[7]Ibid., 209.

[8]Ibid.,247-8.

[9]C. Bernard Ruffin, *Padre Pio: The True Story* (Huntington, IN: Our Sunday Visitor, 2018). 411, 413-4, 419, 421-2.

[10]http://chiesa.espresso.repubblica.it/articolo/7149.html

[11]C. Bernard Ruffin, op cit., 415.

[12]Ibid., 394.

[13]bid.,397-9.

[14]https://caccioppoli.com/Padre%20Pio%2073-81%20(%2760-%2768).html#_ftn2

[15]Ibid., 405.

[16]Ibid.

[17]https://www.amazon.com/Padre-Pio-Miracles-Politics-Secular/product-reviews/0805089055

Ingram Content Group UK Ltd.
Milton Keynes UK
UKHW021309190423
420429UK00022B/714